Architecting and Building High-Speed SoCs

Design, develop, and debug complex FPGA-based systems-on-chip

Mounir Maaref

BIRMINGHAM—MUMBAI

Architecting and Building High-Speed SoCs

Group Product Manager: Rahul Nair
Publishing Product Manager: Meeta Rajani
Senior Editor: Athikho Sapuni Rishana
Technical Editor: Rajat Sharma
Copy Editor: Safis Editing
Project Coordinator: Ashwin Kharwa
Proofreader: Safis Editing
Indexer: Rekha Nair
Production Designer: Aparna Bhagat
Marketing Coordinator: Nimisha Dua

First published: December 2022

Production reference: 1111122

Published by Packt Publishing Ltd.
Livery Place
35 Livery Street
Birmingham
B3 2PB, UK.

ISBN 978-1-80181-099-9

www.packt.com

To my very dear parents, Rahma and Kouider, may they accept here the tribute of my gratitude, which, however great it may be, will never equal their tenderness and their devotion. To my beloved wife, Nadia, who fills my life with joy and happiness, and who without her support this work would not have been possible. To my lovely children, Anissa and Rayan, who give me the courage to persevere in my endeavors and the chance to be a parent, they help me every day give meaning to my life. To my dear sisters and brothers and to all my family and friends that they find here the expression of my most sincere feelings.

Contributors

About the author

Mounir Maaref lives in the UK and works as a Principal SoC Architect. He has 25 years' experience in the microelectronics industry spanning FPGAs, ASICs, embedded processing, networking, data storage, satellite communications, Bluetooth, and WiFi connectivity. He likes working on cutting edge technologies involving both hardware and software. His main focus is on the system architecture design, hardware and software interactions, performance analysis, and modeling. He has published several application notes and white papers and has been a speaker at many conferences worldwide. He holds a masters degree in Electronics and Telecoms. He is a 2nd dan black belt in Tang Soo Do and is getting trained to become a martial arts instructor.

I would like to thank all the friendly and professional staff at Packt, with whom I was privileged to interact at all the stages of writing and publishing this work. Their guidance and help were crucial in transforming my ideas into the material that we share in this book. I am immensely grateful to Martin Abrahams, my friend and ex-colleague at Micron, who took the time to review my work and provide valuable feedback that only made this book better. I would also like to thank all my friends and colleagues at Samsung Cambridge Solutions Centre, from whom I have been receiving a great opportunity to learn and contribute.

About the reviewer

Martin Abrahams has worked in the semiconductor industry for over sixteen years in medical and automotive image sensor development for Micron and Onsemi. He is the architect for Onsemi of virtual prototyping for embedded microprocessor development using SystemC transaction-level modeling and leads the development of hardware component models for digital verification using SystemVerilog and UVM. Martin received a first class honours bachelor of science degree from the University of Bristol and was awarded a doctor of philosophy by the University of Southampton. Earlier in his career, he was a founder at EDA startup TransEDA. TransEDA developed the first VHDL and Verilog code coverage tools which are now central to modern RTL verification.

I would like to thank my family and colleagues at Onsemi for the time they have granted me to investigate, develop, and deploy new design and verification methodologies that transform the way we design digital electronics. I must also acknowledge their patience and encouragement through the many attempts it takes to get to a working solution. I would also like to thank Mounir for his kind invitation to review his book.

Table of Contents

2

FPGA Devices and SoC Design Tools   29

3

Basic and Advanced On-Chip Busses and Interconnects   49

4

Connecting High-Speed Devices Using Buses and Interconnects   93

5

Basic and Advanced SoC Interfaces   125

Part 2: Implementing High-Speed SoC Designs in an FPGA

9

SoC Design Hardware and Software Integration   289

Part 3: Implementation and Integration of Advanced High-Speed FPGA SoCs

10

Building a Complex SoC Hardware Targeting an FPGA   311

11

Addressing the Security Aspects of an FPGA-Based SoC   327

12

Building a Complex Software with an Embedded Operating System Flow   341

13

Video, Image, and DSP Processing Principles in an FPGA and SoCs   359

Preface

Modern and complex SoCs can adapt to many demanding system requirements by combining the processing power of ARM processors and the feature-rich Xilinx FPGAs. You'll need to understand many protocols, use a variety of internal and external interfaces, pinpoint the bottlenecks, and define the architecture of an SoC in an FPGA to produce a superior solution in optimal time and with an optimal cost. This book adopts a practical approach to help you master both the hardware and software design flows, understand key interconnects and interfaces, analyze the system performance and enhance it using acceleration techniques, and finally, build an RTOS-based software application for an advanced SoC design.

Beginning with an introduction to the FPGA SoC technology fundamentals and their associated development design tools, this book will guide you in building the SoC hardware and software, starting from the architecture definition to testing on a demo board or a virtual platform. The level of complexity evolves as the book progresses and it covers advanced applications such as communications, security, and coherent hardware acceleration.

By the end of this book, you'll have learned the concepts underlying FPGA SoC advanced features and you'll have constructed a high-speed SoC targeting a high-end FPGA from the ground up.

Who this book is for

This book is intended for FPGA and ASIC hardware and firmware developers, IoT engineers, SoC architects, and anyone interested in understanding the process of developing a complex SoC, including all aspects of the hardware design and the associated firmware design. Prior knowledge of digital electronics and some experience in coding in **Very High-Speed Design Language** (**VHDL**) or Verilog and C, or a similar language suitable for embedded systems, will be required for using this book. A general understanding of FPGA and CPU architecture will be helpful but not mandatory for using this book.

What this book covers

Chapter 1, *Introducing FPGA Devices and SoCs*, begins by describing the FPGA technology and its evolution since it was first invented by Xilinx in the 1980s. It goes over the electronics industry gap that the FPGA devices cover, their adoption, and their ease of use to implement custom digital hardware functions and systems. It then describes the high-speed SoCs and their evolution since they were introduced as a solution by the major FPGA vendors in the early 2000s. It looks at SoC classification for the targeted applications, specifically for FPGA implementations.

Chapter 2, FPGA Devices and SoC Design Tools, begins by giving an overview of the Xilinx FPGA hardware design flow in general and the tools associated with it. It then highlights the specific tools used when designing an SoC for FPGAs. It also introduces SoC design hardware verification using the available simulation tools. The chapter also covers the software design flow and its different steps and introduces the tools involved in every step of the software design for an FPGA-based SoC.

Chapter 3, Basic and Advanced On-Chip Busses and Interconnects, begins by giving an overview of the busses and interconnects used within an SoC. It introduces the concepts of data sharing and coherency and how to solve their associated challenges. It gives a good introduction to the AMBA and OCP protocols. It also covers data movement within an SoC and the use of DMA engines.

Chapter 4, Connecting High-Speed Devices Using Busses and Interconnects, begins by giving an overview of the busses and interconnects used off-chip to connect an SoC and/or an FPGA to other high-speed devices on the electronics board. It introduces the PCIe interconnect, the Ethernet interconnect, and the emerging Gen-Z protocol. It also introduces the emerging CCIX interconnect protocol and the concept of extending data coherency off-chip by adding protocol layers to manage it.

Chapter 5, Basic and Advanced SoC Interfaces, begins by defining an SoC interface for a given function. It classifies the SoC interfaces and lists their associated controller services. Then, the chapter covers processor caches and their organizations with a focus on ARMv7 architecture. It also introduces the processor memory management unit and its role in virtual-to-physical address translation and in implementing address space management and protection. It delves into the different memory and storage interfaces for on-chip and off-chip memories, their topologies and architectural features, and the criteria for choosing a given interface (or a combination of many).

Chapter 6, What Goes Where in a High-Speed SoC Design, teaches you about the SoC architecture definition phase that precedes the design and implementation phases. This phase is very useful to system architects as it translates a certain set of product requirements into a high-level description of the SoC design to accomplish. It details the criteria used during the functional decomposition stage in which a trade-off is reached between what is better suited to be implemented in hardware and what is rather a good target for a software implementation. It gives an overview of SoC system modeling using many available tools and environments.

Chapter 7, FPGA SoC Hardware Design and Verification Flow, delves into building the SoC hardware using all the tools introduced in the previous chapters. This chapter is hands-on, where you will build a simple but complete SoC for a Xilinx FPGA. You are guided through every step of the SoC hardware design phases, from the concept to the FPGA image generation. The chapter will also cover the hardware verification aspects, such as using the available **Register Transfer Level** (RTL) simulation tools to simulate part of the design and check for potential hardware issues.

Chapter 8, FPGA SoC Software Design Flow, focuses on the steps involved in building the software that will run on the SoC processors. You will first configure the software components needed by this phase of the design process, such as customizing the **Board Support Package** (BSP), configuring the libraries, and customizing the drivers for a simple application. You will revisit the SoC project built in the previous chapter to learn how to define a distributed software microarchitecture and will go through the steps of building all the project software components using bare-metal software applications targeting the SoC hardware.

Chapter 9, SoC Design Hardware and Software Integration, helps you to download an FPGA binary configuration file to the device and boot the SoC CPU's phase or target an emulation platform if a demo board isn't available. You will debug the software running on the target platform (real hardware or virtual models) and gain practical familiarity with the available software debugging tools. You will also learn how to evaluate the software performance and understand its associated metrics using the software profiling tools in order to highlight any areas of concern in the designed system.

Chapter 10, Building a Complex SoC Hardware Targeting an FPGA, introduces you to some of the SoC design advanced topics that present many challenges to design engineers given their multidimensional nature. It will continue with the same practical approach as previous chapters by first adding more complex elements to the hardware design. It will now be built to host an embedded operating system as well. You will be introduced to the hardware acceleration techniques to help augment the system performance and equipped with the fundamental knowledge to make this step challenge-free. You will examine the different ways they can be applied and what system aspects need to be considered at the architectural level in the shared data paradigm.

Chapter 11, Addressing the Security Aspects of an FPGA-Based SoC, introduces you to the SoC security aspects and how these aspects are addressed by the FPGA SoC hardware. You will then learn about the security paradigms available in the ARM-based processors within the SoC hardware. The chapter will then introduce the security aspects from a software perspective and how they make use of the previously mentioned hardware security features to build a secure SoC in an FPGA.

Chapter 12, Building a Complex Software with an Embedded Operating System Flow, teaches you about the flow and helps you discover the tools used to build a complex software application to run on the complex FPGA SoC. You will use the design tools available to create the SoC BSP for the targeted embedded operating system, such as FreeRTOS. You will go through the process of generating an embedded bootloader for the target application to be used at runtime when the SoC is powered up or reset.

Chapter 13, Video, Image, and DSP Processing Principles in an FPGA and SoCs, introduces some of the advanced applications implemented in modern FPGAs and SoCs and what makes these devices such powerful compute engines for these types of compute- and bandwidth-demanding applications. It will clarify how parallel processing required by DSP applications in general can be easily implemented in the FPGA logic and how these parallel compute engines can be interfaced internally and externally to wide memories and internally to the powerful CPUs available in the SoCs.

Chapter 14, Communication and Control System Implementation in FPGAs and SoCs, continues introducing more advanced applications implemented in modern FPGAs and SoCs and explains what makes these devices such powerful compute engines for these types of I/O- and bandwidth-demanding applications. It will focus on some of the communication protocols that can make use of the FPGA multi-Gb transceivers, the logic that can perform packet inspections and filtering, and the CPU that can implement algorithms in the SW to manage the communication stack and interface to the user and other onboard devices. It will also cover control applications in the FPGA and SoCs and how they can benefit from all of their available features.

To get the most out of this book

You need to have familiarity with digital electronics in general and, specifically, you need to have some fundamental knowledge of modern logic design at a RTL using a hardware design language such as VHDL, Verilog, or SystemVerilog. You will also need some working knowledge of embedded programming using a high-level language such as C or C++ and have some experience using cross-compilers to build executables for a target embedded processor. The hardware design flow and the embedded software design flow both use tools packaged within the Xilinx Vivado and Vitis environments. You will be guided through their installation processes on your host machine. If you are running a Windows operating system on your host machine, you will be guided through installing VirtualBox, which is an Oracle hypervisor to host a Linux guest operating system to be used as your development host operating system.

Software/hardware covered in the book	Operating system requirements
Xilinx Vivado 2021.2 or a higher version	Windows Enterprise and Professional 10.0
	Or Window 10 HE and VirtualBox
	Or Ubuntu 16.04.5 LTS up to 20.04.1 LTS
Xilinx Vitis	Windows Enterprise and Professional 10.0
	Or Window 10 HE and VirtualBox
	Or Ubuntu 16.04.5 LTS up to 20.04.1 LTS
VirtualBox	Windows 10

All the required installation steps are described in detail in the book using a simple, logical step-by-step approach that will get you up and running with the tools. Simply follow the instructions and don't skip any steps during the installation process. Also, make sure you perform the required configuration when indicated to do so, so the tools are ready to use without wasting time debugging issues that are sometimes hard to track if a step in the installation or configuration process has been skipped or omitted.

If you are using the digital version of this book, we advise you to type the code yourself or access the code from the book's GitHub repository (a link is available in the next section). Doing so will help you avoid any potential errors related to the copying and pasting of code.

If you can get hold of a Xilinx Zynq-7000 SoC board, this will be great as you can download the FPGA bitstream and the executable software to the board and perform debugging and profiling on the real hardware. Nevertheless, if you don't have access to a Xilinx Zynq-7000 SoC board, you can still use a virtual target within the Xilinx tools to debug and interact with the executable software running on the virtual target.

Download the example code files

You can download the example code files for this book from GitHub at `https://github.com/PacktPublishing/Architecting-and-Building-High-Speed-SoCs`. If there's an update to the code, it will be updated in the GitHub repository.

We also have other code bundles from our rich catalog of books and videos available at `https://github.com/PacktPublishing/`. Check them out!

Code in Action

The Code in Action videos for this book can be viewed at `http://bit.ly/3NNFkZs`.

Download the color images

We also provide a PDF file that has color images of the screenshots and diagrams used in this book. You can download it here: `https://packt.link/Hjk2H`.

Conventions used

There are a number of text conventions used throughout this book.

`Code in text`: Indicates code words in text, database table names, folder names, filenames, file extensions, pathnames, dummy URLs, user input, and Twitter handles. Here is an example: "In PM, Vivado Design Suite uses a project file (`.xpr`) and directory structure to maintain the design source files."

Bold: Indicates a new term, an important word, or words that you see onscreen. For instance, words in menus or dialog boxes appear in **bold**. Here is an example: "You first need to launch the Vivado IDE, then on the Vivado launch screen, choose **Quick Start**, then **Create Project**."

> **Tips or important notes**
> Appear like this.

Get in touch

Feedback from our readers is always welcome.

General feedback: If you have questions about any aspect of this book, email us at `customercare@packtpub.com` and mention the book title in the subject of your message.

Errata: Although we have taken every care to ensure the accuracy of our content, mistakes do happen. If you have found a mistake in this book, we would be grateful if you would report this to us. Please visit `www.packtpub.com/support/errata` and fill in the form.

Piracy: If you come across any illegal copies of our works in any form on the internet, we would be grateful if you would provide us with the location address or website name. Please contact us at `copyright@packt.com` with a link to the material.

If you are interested in becoming an author: If there is a topic that you have expertise in and you are interested in either writing or contributing to a book, please visit `authors.packtpub.com`.

Share Your Thoughts

Once you've read *Architecting and Building High Speed SoCs*, we'd love to hear your thoughts! Scan the QR code below to go straight to the Amazon review page for this book and share your feedback.

`https://packt.link/r/1801810990`

Your review is important to us and the tech community and will help us make sure we're delivering excellent quality content.

Download a free PDF copy of this book

Thanks for purchasing this book!

Do you like to read on the go but are unable to carry your print books everywhere?

Is your eBook purchase not compatible with the device of your choice?

Don't worry, now with every Packt book you get a DRM-free PDF version of that book at no cost.

Read anywhere, any place, on any device. Search, copy, and paste code from your favorite technical books directly into your application.

The perks don't stop there, you can get exclusive access to discounts, newsletters, and great free content in your inbox daily

Follow these simple steps to get the benefits:

1. Scan the QR code or visit the link below

https://packt.link/free-ebook/9781801810999

2. Submit your proof of purchase
3. That's it! We'll send your free PDF and other benefits to your email directly

Part 1: Fundamentals and the Main Features of High-Speed SoC and FPGA Designs

This part introduces the main features and building blocks of SoCs and FPGA devices and associated design tools and provides an overview of the main on-chip and off-chip interconnects and interfaces.

This part comprises the following chapters:

- *Chapter 1, Introducing FPGA Devices and SoCs*
- *Chapter 2, FPGA Devices and SoC Design Tools*
- *Chapter 3, Basic and Advanced On-Chip Busses and Interconnects*
- *Chapter 4, Connecting High-Speed Devices Using Busses and Interconnects*
- *Chapter 5, Basic and Advanced SoC Interfaces*

1
Introducing FPGA Devices and SoCs

In this chapter, we will begin by describing what the **field-programmable gate array** (**FPGA**) technology is and its evolution since it was first invented by Xilinx in the 1980s. We will cover the electronics industry gap that FPGA devices cover, their adoption, and their ease of use for implementing custom digital hardware functions and systems. Then, we will describe the high-speed FPGA-based **system-on-a-chip** (**SoC**) and its evolution since it was introduced as a solution by the major FPGA vendors in the early 2000s. Finally, we will look at how various applications classify SoCs, specifically for FPGA implementations.

In this chapter, we're going to cover the following main topics:

- Xilinx FPGA devices overview
- Xilinx SoC overview and history
- Xilinx Zynq-7000 SoC family hardware features
- Xilinx Zynq UltraScale+ MPSoC family hardware features
- SoC in ASIC technologies

Xilinx FPGA devices overview

An FPGA is a **very large-scale integration** (**VLSI**) **integrated circuit** (**IC**) that can contain hundreds of thousands of **configurable logic blocks** (**CLBs**), tens of thousands of predefined hardware functional blocks, hundreds of predefined external interfaces, thousands of memory blocks, thousands of **input/output** (**I/O**) pads, and even a fully predefined SoC centered around an IBM PowerPC or an ARM Cortex-A class processor in certain FPGA families. These functional elements are optimally spread around the FPGA silicon area and can be interconnected via programmable routing resources. This allows them to behave in a functional manner that's desired by a logic designer so that they can meet certain design specifications and product requirements.

Application-specific integrated circuits (ASICs) and **application-specific standard products (ASSPs)** are VLSI devices that have been architected, designed, and implemented for a given product or a particular application domain. In contrast to ASICs and ASSPs, FPGA devices are generic ICs that can be programmed to be used in many applications and industries. FPGAs are usually reprogrammable as they are based on **static random-access memory (SRAM)** technology, but there is a type that is only programmed once: **one-time programmable (OTP)** FPGAs. Standard SRAM-based FPGAs can be reprogrammed as their design evolves or changes, even once they have been populated in the electronics design board and after being deployed in the field. The following diagram illustrates the concept of an FPGA IC:

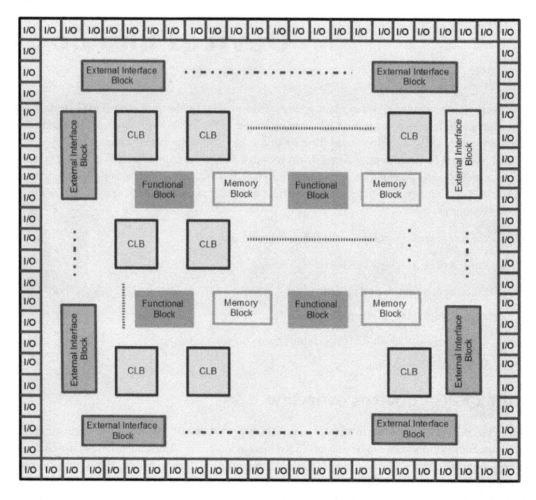

Figure 1.1 – FPGA IC conceptual diagram

As we can see, the FPGA device is structured as a pool of resources that the design assembles to perform a given logical task.

Once the FPGA's design has been finalized, a corresponding configuration binary file is generated to program the FPGA device. This is typically done directly from the host machine at development and verification time over JTAG. Alternatively, the configuration file can be stored in a non-volatile media on the electronics board and used to program the FPGA at powerup.

A brief historical overview

Xilinx shipped its first FPGA in 1985 and its first device was the XC2064; it offered 800 gates and was produced on a 2.0µ process. The Virtex UltraScale+ FPGAs, some of the latest Xilinx devices, are produced in a 14nm process node and offer high performance and a dense integration capability. Some modern FPGAs use 3D ICs **stacked silicon interconnect (SSI)** technology to work around the limitations of Moore's law and pack multiple dies within the same package. Consequently, they now provide an immense 9 million system logic cells in a single FPGA device, a four order of magnitude increase in capacity alone compared to the first FPGA; that is, XC2064. Modern FPGAs have also evolved in terms of their functionality, higher external interface bandwidth, and a vast choice of supported I/O standards. Since their initial inception, the industry has seen a multitude of quantitative and qualitative advances in FPGA devices' performance, density, and integrated functionalities. Also, the adoption of the technology has seen a major evolution, which has been aided by adequate pricing and Moore's law advancements. These breakthroughs, combined with matching advances in software development tools, **intellectual property (IP)**, and support technologies, have created a revolution in logic design that has also penetrated the SoC segment.

There has also been the emergence of the new Xilinx Versal devices portfolio, which targets the data center's workload acceleration and offers a new AI-oriented architecture. This device class family is outside the scope of this book.

FPGA devices and penetrated vertical markets

FPGAs were initially used as the electronics board *glue logic* of digital devices. They were used to implement buses, decode functions, and patch minor issues discovered in the board ASICs post-production. This was due to their limited capacities and functionalities. Today's FPGAs can be used as the hearts of smart systems and are designed with their full capacities in terms of parallel processing and their flexible adaptability to emerging and changing standards, specifically at the higher layers, such as the Link and Transactions layers of new communication or interface protocols. These make reconfiguring FPGA the obvious choice in medium or even large deployments of these emerging systems. With the addition of ASIC class embedded processing platforms within the FPGA for integrating a full SoC, FPGA applications have expanded even deeper into industry verticals where it has seen limited useability in the past. It is also very clear that, with the prohibitive cost of **non-recurring engineering (NRE)** and producing ASICs at the current process nodes, FPGAs are becoming the first choice for certain applications. They also offer a very short time to market for certain segments where such a factor is critical for the product's success.

FPGAs can be found across the board in the high-tech sector and range from the classical fields such as wired and wireless communication, networking, defense, aerospace, industrial, **audio-video broadcast** (**AVB**), ASIC prototyping, instrumentation, and medical verticals to the modern era of ADAS, data centers, the cloud and edge computing, **high-performance computing** (**HPC**), and ASIC emulation simulators. They have an appealing reason to be used almost everywhere in an electronics-based application.

An overview of the Xilinx FPGA device families

Xilinx provides a comprehensive portfolio of FPGA devices to address different system design requirements across a wide range of the application's spectrum. For example, Xilinx FPGA devices can help system designers construct a base platform for a high-performance networking application necessitating a very dense logic capacity, a very wide bandwidth, and performance. They can also be used for low-cost, small-footprint logic design applications using one of the low-cost FPGA devices either for high or low-volume end applications.

In this large offering, there are the cost-optimized families such as the Spartan-7 family and the Spartan-6 family, which are built using a 45nm process node, the Artix-7 family, and the Zynq-7000 family, which is built using a 28nm process node.

There is also the 7-series family in a 28nm process, which includes the Artix-7, Kintex-7, and Virtex-7 families of FPGAs, in addition to the Spartan-7 family.

Additionally, there are FPGAs from the UltraScale Kintex and Virtex families in a 20nm process node.

The UltraScale+ category contains three more additional families – the Artix UltraScale+, the Kintex UltraScale+, and the Virtex UltraScale+, all in a 16nm process node.

Each device family has a matrixial offering table that is defined by the density of logic, the number of functional hardware blocks, the capacity of the internal memory blocks, and the amount of I/Os in each package. This makes the offered combinations an interesting catalog to pick a device that meets the requirements of the system to build using the specific FPGA. To examine a given device offering matrix, you need to consult the specific FPGA family product table and product selection guide. For example, for the UltraScale+ FPGAs, please go to `https://www.xilinx.com/content/dam/xilinx/support/documentation/selection-guides/ultrascale-plus-fpga-product-selection-guide.pdf`.

An overview of the Xilinx FPGA devices features

As highlighted in the introduction to this chapter, modern Xilinx FPGA devices contain a vast list of hardware block features and external interfaces that relatively define their category or family and, consequently, make them suitable for a certain application or a specific market vertical. This chapter looks at the rich list of these features to help you understand what today's FPGAs are capable of offering system designers. It is worth noting that not all the FPGAs contain all these elements.

For a detailed overview of these features, you are encouraged to examine the Xilinx UltraScale+ Data Sheet as a good starting point at `https://www.xilinx.com/content/dam/xilinx/support/documentation/data_sheets/ds890-ultrascale-overview.pdf`.

In the following subsections, we will summarize some of these features.

Logic elements

Modern Xilinx FPGAs have an abundance of CLBs. These CLBs are formed by **lookup tables (LUTs)** and registers known as **flip-flops**. These CLBs are the elementary ingredients that logic user functions are built from to form the desired engine to perform a combinatorial function that's coupled (or not) with sequential logic. These are also built from Flip-Flop resources contained within the CLBs. Following a full design process from design capture, to synthesizing and implementing the production of a binary image to program the FPGA device, these CLBs are configured to operate in a manner that matches the aforementioned required partial task within the desired function defined by the user. The CLB can also be configured to behave as a deep shift register, a multiplexer, or a carry logic function. It can also be configured as distributed memory from which more SRAM memory is synthesized to complement the SRAM resources that can be built using the FPGA device block's RAM.

Storage

Xilinx FPGAs have many block RAMs with built-in FIFO. Additionally, in UltraScale+ devices, there are 4Kx72 UltraRAM blocks. As mentioned previously, the CLB can also be configured as distributed memory from which more SRAM memory can be synthesized.

The Virtex UltraScale+ HBM FPGAs can integrate up to 16 GB of **high-bandwidth memory (HBM)** Gen2.

Xilinx Zynq UltraScale+ MPSoC also provides many layers of SRAM memory within its ARM-based SoC, such as OCM memory and the Level 1 and Level 2 caches of the integrated CPUs and GPUs.

Signal processing

Xilinx FPGAs are rich in resources for **digital signal processing** (DSP). They have DSP slices with 27x18 multipliers and rich local interconnects. The DSP slice has many usage possibilities, as described in the FPGA datasheet.

Routing and SSI

The Xilinx FPGA's device interconnect employs a routing infrastructure, which is a combination of configurable switches and nets. These allow the FPGA elements such as the I/O blocks, the DSP slices, the memories, and the CLBs to be interconnected.

The efficiency of using these routing resources is as important as the device hardware's logical resources and features. This is because they represent the nerve system of the FPGA device, their abundance of interconnect logic, and their functional elements, which are crucial to meeting the design performance criteria.

Design clocking

Xilinx FPGA devices contain many clock management elements, including **digital local loops** (**DLLs**) for clock generation and synthesis, global buffers for clock signal buffering, and routing infrastructure to meet the demands of many challenging design requirements. The flexibility of the clocking network minimizes the inter-signal delays or skews.

External memory interfaces

The Xilinx FPGAs can interface to many external parallel memories, including DDR4 SDRAM. Some FPGAs also support interfacing to external serial memories, such as **Hybrid Memory Cube** (**HMC**).

External interfaces

Xilinx FPGA devices interface to the external ICs through I/Os that support many standards and PHY protocols, including the serial **multi-gigabit transceivers** (**MGTs**), Ethernet, PCIe, and Interlaken.

ARM-based processing subsystem

The first device family that Xilinx brought to the market that integrated an ARM CPU was the Zynq-7000 SoC FPGA with its integrated ARM Cortex-A9 CPU. This family was followed by the Xilinx Zynq UltraScale+ MPSoCs and RFSoCs, which feature a **processing system** (**PS**) that includes a dual or a quad-core variant of the ARM Cortex-A53, and a dual-core ARM Cortex-R5F. Some variants have a **graphics processing unit** (**GPU**). We will delve into the Xilinx SoCs in the next chapter.

Configuration and system monitoring

Being SRAM-based, the FPGA requires a configuration file to be loaded when powered up to define its functionality. Consequently, any errors that are encountered in the FPGA's configuration binary image, either at configuration time or because of a physical problem in mission mode, will alter the overall system functionality and may even cause a disastrous outcome for sensitive applications. Therefore, it is a necessity for critical applications to have system monitoring to urgently intervene when such an error is discovered to correct it and limit any potential damage via its built-in self-monitoring mechanism.

Encryption

Modern FPGAs provide decryption blocks to address security needs and protect the device's hardware from hacking. FPGAs with integrated SoC and PS blocks have a **configuration and security unit (CSU)** that allows the device to be booted and configured safely.

Xilinx SoC overview and history

In the early 2000s, Xilinx introduced the concept of building embedded processors into its available FPGAs at the time, namely the Spartan-2, Virtex-II, and Virtex-II Pro families. Xilinx brought two flavors of these early SoCs to the market: a soft version and an initial hard macro-based option in the Virtex-II Pro FPGAs.

The soft flavor uses MicroBlaze, a Xilinx RISC 32-bit based soft processor coupled initially with an IBM-based bus infrastructure called CoreConnect and a rich set of peripherals, such as a Gigabits Ethernet MACs, PCIe, and DDR DRAM, just to name a few. A typical MicroBlaze soft processor-based SoC looks as follows:

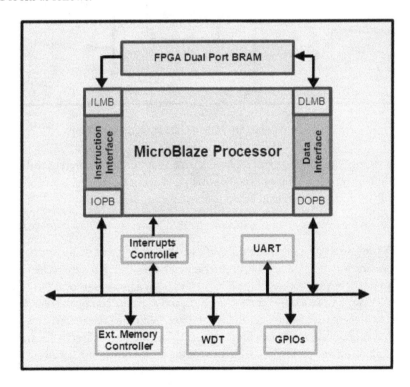

Figure 1.2 – Legacy FPGA MicroBlaze embedded system

The hard macro version uses a 32-bit IBM PowerPC 405 processor. It includes the CPU core, a **memory management unit** (**MMU**), 16 KB L1 data and 16 KB L1 instruction caches, timer resources, the necessary debug and trace interfaces, the CPU CoreConnect-based interfaces, and a fast memory interface known as **on-chip memory** (**OCM**). The OCM connects to a mapped region of internal SRAM that's been built using the FPGA block RAMs for fast code and data access. The following diagram shows a PowerPC 405 embedded system in a Virtex-II Pro FPGA device:

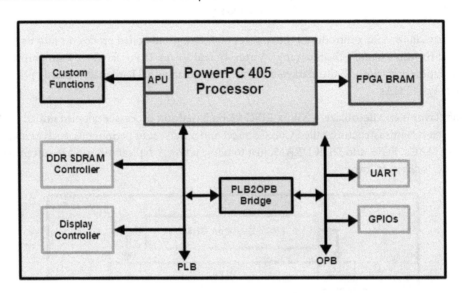

Figure 1.3 – Virtex-II Pro PowerPC405 embedded system

Embedded processing within FPGAs has received a wide adoption from different vertical spaces and opened the path to many single-chip applications that previously required the use of an external CPU, alongside the FPGA device, as the main board processor.

The Virtex-4 FX was the next generation to include the IBM PowerPC 405 and improved its core speed.

The Virtex-5 FXT followed and integrated the IBM PowerPC 440x5 CPU, a dual-issue superscalar 32-bit embedded processor with an MMU, a 32 KB instruction cache, a 32 KB data cache, and a Crossbar interconnect. To interface with the rest of the FPGA logic, it has a **processor local bus** (**PLB**) interface, an **auxiliary processor unit** (**APU**) for connecting FPU, and a custom coprocessor built into the FPAG logic. It also has a high-speed memory controller interface. With the Ethernet Tri-Speed 10/100/1000 MACs integrated as hardware functional blocks in the FPGA, we started seeing the main ingredients necessary for making an SoC in FPGAs, with most of the logic-consuming hardware functions now bundled together around the CPU block or delivered as a hardware functional block that just needs interfacing and connecting to the CPU. This was a step close to a full SoC in FPGAs. The following diagram shows a PowerPC 440 embedded system in a Virtex-5 FXT FPGA device:

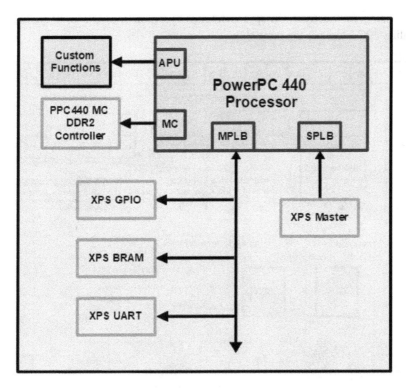

Figure 1.4 – Virtex-5 FXT PowerPC440 embedded system

The Virtex-5 FXT was the last Xilinx FPGA to include an IBM-based CPU; the future was switching to ARM and providing a full SoC in FPGAs with the possibility to interface to the FPGA logic through adequate ports. This offered the industry a new kind of SoC that, within the same device, combined the power of an ASIC and the programmability of the Xilinx-rich FPGAs. This brings us to this book's main topic, where we will delve into and try to deal with all Xilinx's related design development and technological aspects while taking an easy-to-follow and progressive approach.

The following diagram illustrates the approach taken by Xilinx to couple an ARM-based CPU SoC with the Xilinx FPGA logic in the same chip:

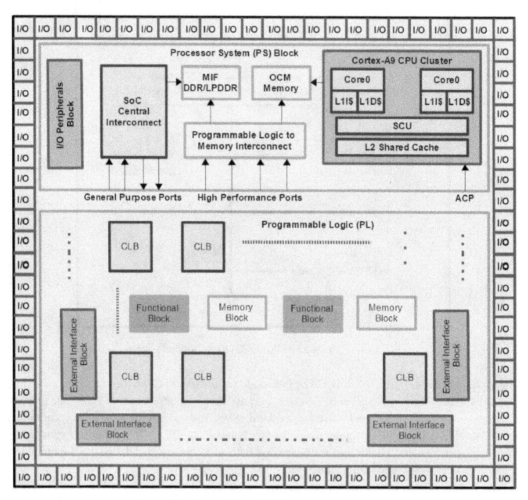

Figure 1.5 – Zynq-7000 SoC FPGA conceptual diagram

A short survey of the Xilinx SoC FPGAs based on an ARM CPU

The first device family that Xilinx brought to the market for integrating an ARM Cortex-A9 CPU was the Zynq-7000 FPGA. The Cortex-A9 is a 32-bit processor that implements the ARMv7-A architecture and can run many instruction formats. These are available in two configurations: a single Cortex-A9 core in the Zynq-7000S devices and a dual Cortex-A9 cluster in the Zynq-7000 devices.

The next generation that followed was the Zynq UltraScale+ MPSoC devices, which provide a 64-bit ARM CPU cluster for integrating an ARM Cortex-A53, coupled with a 32-bit ARM Cortex-R5 in the same SoC. The Cortex-A53 CPU implements the ARMv8-A architecture, while the Cortex-R5 implements the ARMv7 architecture and, specifically, the R profile. The Zynq UltraScale+ MPSoC comes in different configurations. There is the CG series with a dual-core Cortex-A53 cluster, the EG series with a quad-core Cortex-A53 cluster and an ARM MALI GPU, and the EV series, which comes with additional video codecs to what is available in the EG series.

A few years ago, Xilinx also launched a version of the MPSoC with key components to help build advanced radio connectivity SoCs: the Zynq UltraScale+ RFSoC.

Xilinx Zynq-7000 SoC family hardware features

As mentioned previously, the Zynq FPGA SoC integrates a popular ARM CPU based on the ARMv7, and the classical FPGA part based on the Xilinx 7th generation logic with rich hardware features.

For a detailed description of the Zynq-7000 SoC FPGA and its features, please refer to the SoC **Technical Reference Manual** (**TRM**) available at `https://www.xilinx.com/support/documentation/user_guides/ug585-Zynq-7000-TRM.pdf`.

This section specifies the main Zynq-7000 SoC features and defines them to help you quickly visualize the device's capabilities.

The SoC is mainly composed of an **application processor unit** (**APU**), a connectivity matrix, an OCM memory interface, external memory interfaces, and the **I/O peripherals** (**IOP**) block.

The following diagram provides a detailed architectural view of the Zynq-7000 SoC:

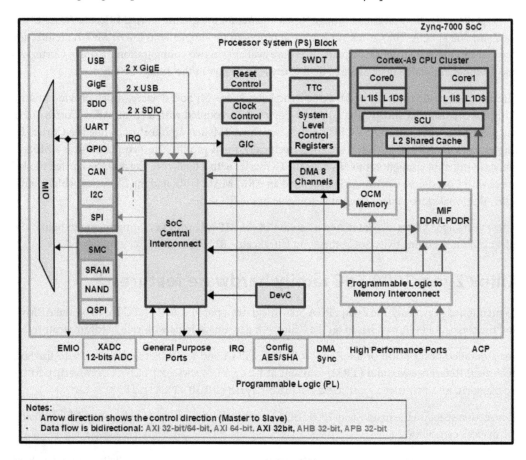

Figure 1.6 – Zynq-7000 SoC architecture – dual-core cluster example

Zynq-7000 SoC APU

The CPU cluster topology is built around an ARM Cortex-A9 CPU, which comes in a dual-core or a single-core MPCore. Each CPU core has an L1 instruction cache and an L1 data cache. It also has its own MMU, a **floating-point unit** (**FPU**), and a NEON SIMD engine. The CPU cluster has an L2 common cache and a **snoop control unit** (**SCU**). This SCU provides an **accelerator coherency port** (**ACP**) that extends cache coherency beyond the cluster with external masters when implemented in the FPGA logic.

Each core provides a performance figure of 2.5 DMIPS/MHz with an operating frequency ranging from 667 MHz to 1 GHz, depending on the Zynq FPGA speed grade. The FPU supports both single and double precision operands with a performance figure of 2.0 MFLOPS/MHz. The CPU core is TrustZone-enabled for secure operation. It supports code compression via the Thumb-2 instructions set. The Level 1 instructions and data caches are both 32 KB in size and are 4-way set-associative.

The CPU cluster supports both SMP and AMP operation modes. The Level 2 cache is 512 KB in size and is common to both CPU cores and for both instructions and data. The L2 cache is an eight-way set associative. The cluster also has a 256 KB OCM RAM that can be accessed by the APU and the **programmable logic (PL)**.

The PS has 8-channel DMA engines that support transactions between memories, peripherals, and scatter-gather operations. Their interfaces are based on the AXI protocol. The FPGA PL can use up to four DMA channels.

The SoC has a **general interrupt controller (GIC)** version 1.0 (GIC v1). The GIC distributes interrupts to the CPU cluster cores according to the user's configuration and provides support for priority and preemption.

The PS supports debugging and tracing and is based on ARM CoreSight interface technology.

Zynq-7000 SoC memory controllers

The Zynq device supports both SDRAM DDR memory and static memories. DDR3/3L/2 and LPDDR2 speeds are supported. The static memory controllers interface to QSPI flash, NAND, and parallel NOR flash.

The SDRAM DDR interface

The SDRAM DDR interface has a dedicated 1 GB of system address space. It can be configured to interface to a full-width 32-bit wide memory or a half-width 16-bit wide memory. It provides support for many DDR protocols. The PS also includes the DDR PHY and can operate at many speeds – up to a maximum of 1,333 Mb/s. This is a multi-port controller that can share the SDRAM DDR memory bandwidth with many SoC clients within the PS or PL regions over four ports. The CPU cluster is connected to a port; two ports serve the PL, while the fourth port is exposed to the SoC central switches, making access possible to all the connected masters.

The following diagram is a memory-centric representation of the SDRAM DDR interface of the Zynq-7000 SoC:

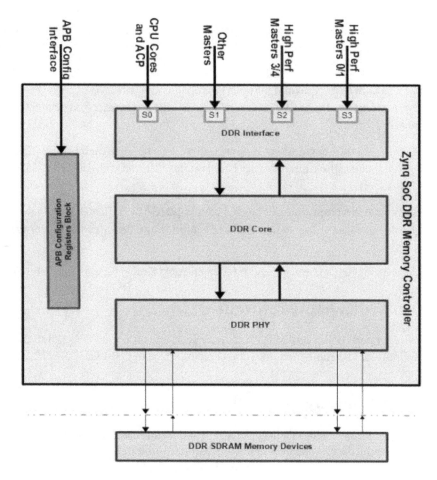

Figure 1.7 – Zynq-7000 SoC DDR SDRAM memory controller

Static memory interfaces

The **static memory controller** (SMC) is based on ARM's PL353 IP. It can interface to NAND flash, SRAM, or NOR flash memories. It can be configured through an APB interface via its operational registers. The SMC supports the following external static memories:

- 64 MB of SRAM in 8-bit width
- 64 MB of parallel NOR flash in 8-bit width
- NAND flash

The following diagram provides a micro-architectural view of the Zynq-7000 SoC SMC:

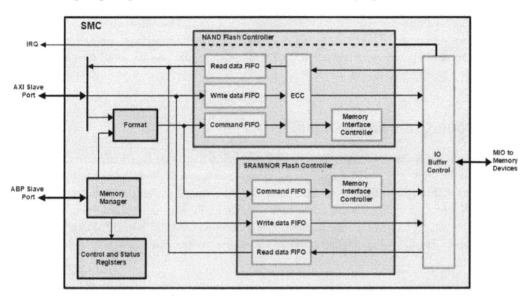

Figure 1.8 – Zynq-7000 SoC static memory controller architecture

QSPI flash controller

The IOP block of the Zynq-7000 SoC includes a QSPI flash interface. It supports serial flash memory devices, as well as three modes of operation: linear addressing mode, I/O mode, and legacy SPI mode.

The software implements the flash device protocol in I/O mode. It provides the commands and data to the controller using the interface registers and reads the received data from the flash memory via the flash registers.

In linear addressing mode, the controller maps the flash address space onto the AXI address space and acts as a translation block between them. Requests that are received on the AXI port of the QSPI controller are converted into the necessary commands and data phases, while read data is put on the AXI bus when it's received from the flash memory device.

In legacy mode, the QSPI interface behaves just like an ordinary SPI controller.

To write the software drivers for a given flash device to control via the Zynq-7000 SoC QSPI controller, you should refer to both the flash device data sheet from the flash vendor and the QSPI controller operational mode settings detailed in the Zynq-7000 TRM. The URL for this was mentioned at the beginning of this section.

The QPSI controller supports multiple flash device arrangements, such as 8-bit access using two parallel devices (to double the device throughput) or a 4-bit dual rank (to increase the memory capacity).

Zynq-7000 I/O peripherals block

The IOP block contains the external communication interfaces and includes two tri-mode (10/100/1 GB) Ethernet MACs, two USB 2.0 OTG peripherals, two full CAN bus interfaces, two SDIO controllers, two full-duplex SPI ports, two high-speed UARTs, and two master and slave I2C interfaces. It also includes four 32-bit banks GPIO. The IOP can interface externally through 54 flexible **multiplexed I/Os (MIOs)**.

Zynq-7000 SoC interconnect

The interconnect is ARM AMBA AXI-based with QoS support. It groups masters and slaves from the PS and extends the connectivity to PL-implemented masters and slaves. Multiple outstanding transactions are supported. Through the Cortex-A9 ACP ports, I/O coherency is possible so that external masters and the CPU cores can coherently share data, minimizing the CPU core cache management operations. The interconnect topology is formed by many switches based on ARM NIC-301 interconnect and AMBA-3 ports. The following diagram provides an overview of the Zynq-7000 SoC interconnect:

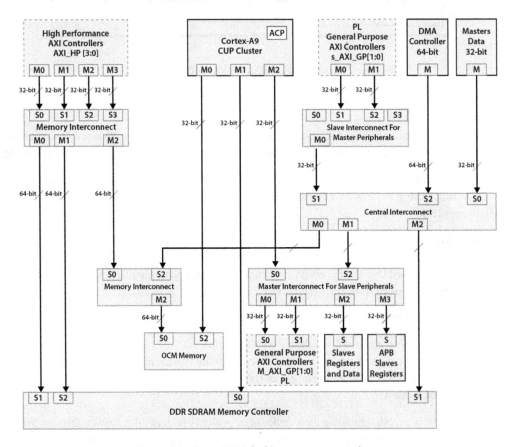

Figure 1.9 – Zynq-7000 SoC interconnect topology

Xilinx Zynq Ultrascale+ MPSoC family overview

The Zynq UltraScale+ MPSoC is the second generation of the Xilinx SoC FPGAs based on the ARM CPU architecture. Like its predecessor, the Zynq-7000 SoC, it is based on the approach of combining the FPGA logic HW configurability and the SW programmability of its ARM CPUs but with improvements in both the FPGA logic and the ARM processor CPUs, as well as its PS features. The UltraScale+ MPSoC offers a heterogeneous topology that couples a powerful 64-bit application processor (implementing the ARMv8-A architecture) and a 32-bit real-time R-profile processor.

The PS includes many types of processing elements: an APU, such as the dual-core or quad-core Cortex-A53 cluster, the dual-core Cortex-R5F **real-time processing unit (RPU)**, the Mali GPU, a PMU, and a **video codec unit (VCU)** in the EG series. The PS has an efficient power management scheme due to its granular power domains control and gated power islands. The Zynq UltraScale+ MPSoC has a configurable system interconnect and offers the user overall flexibility to meet many application requirements. The following diagram provides an architectural view of the Zynq UltraScale+ SoC:

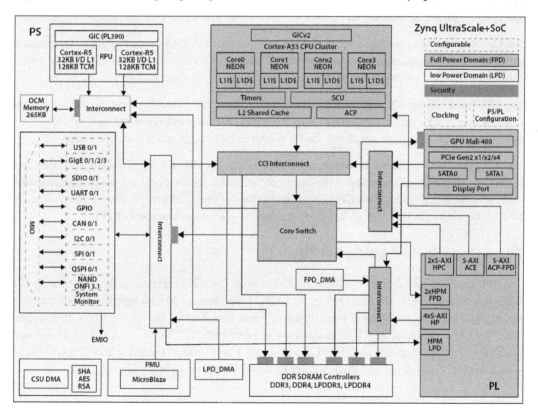

Figure 1.10 – Zynq UltraScale+ MPSoC architecture – quad-core cluster

The following section provides a brief description of the main features of the Zynq UltraScale+ MPSoC. For a detailed technical description, please read the Zynq UltraScale+ MPSoC TRM at `https://www.xilinx.com/support/documentation/user_guides/ug1085-zynq-ultrascale-trm.pdf`.

Zynq UltraScale+ MPSoC APU

The CPU cluster topology is built around an ARM Cortex-A53 CPU, which comes in a quad-core or a dual-core MPCore. The CPU cores implement the Armv8-A architecture with support for the A64 instruction set in AArh64 or the A32/T32 instruction set in AArch32. Each CPU core comes with an L1 instruction cache with parity protection and an L1 data cache with ECC protection. The L1 instruction cache is 2-way set-associative, while the L1 data cache is 4-way set-associative. It also has its own MMU, an FPU, and a Neon SIMD engine. The CPU cluster has a 16-way set-associative L2 common cache and an SCU with an ACP port that extends cache coherency beyond the cluster with external masters in the PL. Each CPU core provides a performance figure of 2.3 DMIPS/MHz with an operating frequency of up to 1.5 GHz. The CPU core is also TrustZone enabled for secure operations.

The CPU cluster can operate in symmetric SMP and asymmetric AMP modes with the power island gating for each processor core. Its unified Level 2 cache is ECC protected, is 1 MB in size, and is common to all CPU cores and both instructions and data.

The APU has a 128-bit **AXI coherent extension** (**ACE**) port that connects to the PS **cache coherent interconnect** (**CCI**), which is associated with the **system memory management unit** (**SMMU**). The APU has an ACP slave port that allows the PL master to coherently access the APU caches.

The APU has a GICv2 **general interrupt controller** (**GIC**). The GIC acts as a distributor of interrupts to the CPU cluster cores according to the user's configuration, with support for priority, preemption, virtualization, and security. Each CPU core contains four of the ARM generic timers. The cluster has a **watchdog timer** (**WDT**), one global timer, and two **triple timers/counters** (**TTCs**).

Zynq UltraScale+ MPSoC RPU

The RPU contains a dual-core ARM Cortex-R5F cluster. The CPU cores are 32-bit real-time profile CPUs based on the ARM-v7R architecture. Each CPU core is associated with **tightly coupled memory** (**TCM**). TCM is deterministic and good for hosting real-time, latency-sensitive application code and data. The CPU cores have 32 KB L1 instruction and data caches. It has an interrupt controller and interfaces to the PS elements and the PL via two AXI-4 ports connected to the low-power domain switch. Software debugging and tracing is done via the ARM CoreSight Debug subsystem.

Zynq UltraScale+ MPSoC GPU

The PS includes an ARM Mali-400 GPU. The GPU includes a **geometry processor (GP)** and has an MMU and a Level 2 cache that's 64 KB in size. The GPU supports OpenGL ES 1.1 and 2.0, as well as OpenVG 1.1 standards.

Zynq UltraScale+ MPSoC VCU

The **video codec unit (VCU)** supports H.265 and H.264 video encoding and decoding standards. The VCU can concurrently encode/decode up to 4Kx2K at 60 **frames per second (FPS)**.

Zynq UltraScale+ MPSoC PMU

The PMU augments the PS with many functionalities for startup and low power modes, some of which are as follows:

- System boot and initialization
- Manages the wakeup events and low processing power tasks when the APU and RPU are in low-power states
- Controls the power-up and restarts on wakeup
- Sequences the low-level events needed for power-up, power-down, and reset
- Manages the clock gating and power domains
- Handles system errors and their associated reporting
- Performs memory scrubbing for error detection at runtime

Zynq UltraScale+ MPSoC DMA channels

The PS has 8-channel DMA engines that support transactions between memories, peripherals, as well as scatter-gather operations. Their interfaces are based on the AXI protocol. They are split into two categories: the **low power domain (LPD)** DMA and **full power domain (FPD)** DMA. The LPD DMA is I/O coherent with the CCI, whereas the FPD DMA is not.

Zynq UltraScale+ MPSoC memory interfaces

In this section, we will look at the various Zynq UltraScale+ MPSoC memory interfaces.

DDR memory controller

The PS has a multiport DDR SDRAM memory controller. Its internal interface consists of six AXI data ports and an AXI control interface. There is a port dedicated to the RPU, while two ports are connected to the CCI; the remaining ports are shared between the DisplayPort controller, the FPD DMA, and the PL. Different types of SDRAM DDR memories are supported, namely DDR3, DDR3L, LPDDR3, DDR4, and LPDDR4.

Static memory interfaces

The external SMC supports managed NAND flash (eMMC 4.51) and NAND flash (24-bit ECC). Serial NOR flash is also supported via 1-bit, 2-bit, Quad-SPI, and dual Quad-SPI (8-bit).

OCM memory

The PS also has an on-chip RAM that's 256 KB in size, which provides low latency storage for the CPU cores. The OCM controller provides eight exclusive access monitors to help implement inter-cluster atomic primitives for access to shared memory regions within the MPSoC.

The OCM memory is implemented as a 32-bit wide memory for achieving a high read/write throughput and uses read-modify-write operations for accesses that are smaller in size. It also has a protection unit and divides the OCM address space into 64 regions, where each region can have separate security and access attributes.

QSPI flash controller

There are two Quad-SPI controllers in the IOP block of the PS, as follows:

- A **legacy Quad-SPI (LQSPI)** controller that presents the flash device as a linear memory space on the AXI interface of the controller. It supports **eXecute-in-Place (XIP)** for booting and running application software.
- A **generic Quad-SPI (GQSPI)** controller that provides I/O, DMA, and SPI mode interfacing. Boot and XIP are not supported by the GQSPI.

The PS can only use a single controller at a time. The Quad-SPI controllers access multi-bit flash memory devices for high throughput and low pin-count applications.

Zynq-UltraScale+ MPSoC IOs

The PS integrates 4-Gb transceivers that can operate at a data rate of up to 6.0 Gb/s. These transceivers can be used as part of the physical layer of the peripherals for high-speed communication.

PCIe interface

The PS includes a PCIe Gen2 with either x1, x2, or x4 width. It can operate as a root complex or endpoint. It can act as a master on its AXI interface using its DMA engine.

SATA interface

The PS integrates two SATA host port interfaces that conform to the SATA 3.1 specification and the **Advanced Host Controller Interface** (**AHCI**) version 1.3. Operation speeds at 1.5 Gb/s, 3.0 Gb/s, and 6.0 Gb/s data rates are supported.

Zynq UltraScale+ MPSoC IOP block

The IOP block contains external communication interfaces. The IOP block includes many external interfaces, such as Ethernet MACs, USB controllers, CAN Bus controllers, SDIO interfaces, SPI and I2C ports, and high-speed UARTs.

Zynq-UltraScale+ MPSoC interconnect

The PS interconnect is formed of multiple switches to connect system resources and is based on the ARM AMBA 4.0. The switches are grouped with high-speed bridges, allowing data and commands to flow freely between them. The PS interconnect has separate segments: a **full-power domain** (**FPD**) and a **low-power domain** (**LPD**). It has QoS and performance monitoring features. It also performs transaction monitoring to avoid interconnect hangs. The interconnect uses the **AXI Isolation Block** (**AIB**) module to isolate ports and allows you to power them down to save power. The interconnect has a CCI-400 to extend cache coherency outside of the APU cluster and an SMMU so that virtual addresses outside of the APU cluster can be used.

SoC in ASIC technologies

Choosing the right SoC to use at the heart of an electronics system is decided based on the system's product requirements in terms of features, performance, production volume, cost, and many other marketing-related metrics and company historical facts. For example, an SoC in an ASIC may be chosen to reduce costs for very high production volumes. Designing an SoC in an ASIC usually has a considerable associated effort and cost compared to an FPGA SoC. It depends on the silicon technology target process node, the functions to include, the packaging, and the overall SoC specification.

This section provides a high-level overview of the SoCs in ASIC technologies and their design flow. This will help you visualize some of the extra design steps and associated costs you need to consider when planning an SoC for an ASIC. There are many other **non-recurring engineering** (**NRE**) costs associated with an ASIC design flow, but covering these is outside the scope of this book. The SoCs in an ASIC hardware design flow provide a good introduction to the SoCs in an FPGA hardware design flow because of their similar principles, although the tools, the target technologies, and the capabilities of each are different.

When designing an SoC for an ASIC process, we must start from a clean sheet and choose the CPU cores to use, the SoC interconnect topology, and the system interfaces, as well as the coprocessors and any hardware IP blocks we need in the SoC to meet the system requirements in terms of performance and power budget. This comes with an associated cost in terms of the design effort, third-party IP licensing fees, as well as production foundry costs.

When using an FPGA, we already have the processing platform architecture decided for us, as we saw with the Zynq-7000 SoC and Zynq UltraScale+ MPSoC. It is their extensibility via the PL and their faster time to market that makes them an attractive option at a certain production volume. Most of the time, we won't make use of all the hardware blocks within the PS in the FPGA SoC since these SoCs are tailored, to a certain extent, to meet many common required features for a specific industry vertical and not a specific end application. However, we don't see this as a big problem if, in terms of power consumption, we can limit it using techniques such as clock and power gating. Some systems may opt to use both options in time, where the systems are deployed using an FPGA SoC, a cost reduction path is provided to move the design to an ASIC as the product matures, and its volume production becomes justifiable for the upfront high cost of an ASIC NRE. This approach is a win-win path where possible.

The SoC design for an ASIC involves putting together the system architecture, which usually contains a collection of components and/or subsystems designed in-house or purchased from a third-party vendor for a licensing fee. These components are interconnected together for the Zynq-7000 SoC or Zynq UltraScale+ MPSoC PS to perform the specified functions. The entire system is built on a single IC that either encapsulates a single silicon die or, as in the latest ASICs, stacks multiple silicon dies interconnected via silicon vias in what is known as **System in a Package** (**SiP**). Like an FPGA SoC, the ASIC categories also include a single or many processors, memories, DSP cores, GPUs, interfaces to external circuitry, I/Os, custom IPs, and Verilog or VHDL modules in the system design.

High-level design steps of an SoC in an ASIC

This section will provide an overview of the different steps involved in designing an ASIC. from the design capture phase to the performance and manufacturability verification step.

Design capture

This is the first design step of an SoC, and it consists of capturing the SoC's specification, partitioning the HW/SW, and selecting the IPs. The design capture could simply be in a text format as an architecture specification document or could be associated with a design capture of the specification in a computer language such as C, C++, SystemC, or SystemVerilog. This design capture isn't necessarily a full SoC system model – it could just be an overall description of the main algorithms and inter-block IPC. However, we can observe the emergence of the usage of full SoC system models by using different environments and fulfilling a diverse set of reasons. Time to market is becoming more of a challenge for many companies that use ASICs because they have to wait for the silicon to be designed and produced, tested, and then assembled with other components on a board to start the software development process. This can take up to a year, assuming that everything runs smoothly. Companies typically use a **virtual prototype (VP)** to help them shorten the system design cycle by around 6 months. Building this VP has an engineering cost and requires many technical skillsets with a need for a deep knowledge of the hardware's architecture and microarchitecture. The following diagram provides an overview of the SoC in ASICs design flow:

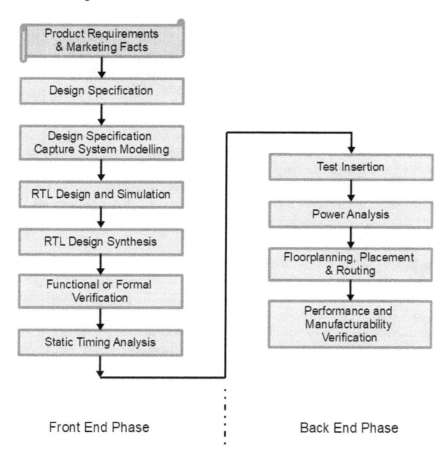

Figure 1.11 – The SoC in ASICs high-level design flow

RTL design

The design capture is followed by the RTL design of the SoC components in an HDL language such as Verilog or VHDL. Then, they are assembled at the top-level module of the SoC. The RTL is then simulated using test benches written specifically to verify the functional correctness – that is, the intended functionality – of the RTL design.

RTL synthesis

Once the RTL design has been completed at a specific module level and simulated using the module verification approach, it is synthesized using a synthesis tool. This step automatically generates a generic gate description from the RTL description. The synthesis tool performs logic optimization for speed and area, which can be guided by the designer via specific scripts or constraints files that are provided alongside the RTL files to the synthesis engine. This step performs state machine decomposition, datapath optimization, and power optimization. Following the extraction and optimization processes, the synthesis tool translates the generic gate-level description into a netlist using a target library. The target library is specific to the ASIC technology process node and foundry.

Functional or formal verification

Following the synthesis step and generating a design netlist, a functional or formal verification step is performed to make sure that there are no residual HDL ambiguities that caused the synthesis tool to produce an incorrect netlist. This step involves rerunning functional verification on the gate-level netlist. Usually, two formal verifications need to be run: model checking, which proves that certain assertions are true, and equivalence checking, which compares two design descriptions.

Static timing analysis

This step verifies the design's timing constraints. It uses a gate delay and routing information to check all the timing paths connecting the logic elements. This requires timing information for any of the IP blocks that are instantiated in the design, such as memories. This analysis will evaluate the timing violations, such as setup and hold times. To ignore any paths or violations forming a special case, the designer can use specific timing constraints to highlight these to the timing analysis tools. This analysis produces a set of results that, for example, report the slack time. The designer uses this information to resynthesize the circuit or redesign it to improve the timing delays in the critical paths.

Test insertion

In this step, various **design for test** (**DFT**) features are inserted. The DFT allows the device to be tested using **automated test equipment** (**ATE**) when the chip is back from the foundry. It consists of many scan-enabled flip-flops and scan chains. There are also **built-in self-test** (**BIST**) blocks **memory built-in self-test** (**MBIST**) blocks, which can apply many testing algorithms to verify the correct functionality of the memories. The Boundary-Scan/JTAG is also added to enable board/system-level testing.

Power analysis

Power analysis tools are used to evaluate the power consumption of the ASIC device. These analyses are statistical and use load models that translate into activity factors for the power consumption estimation.

Floorplanning, placement, and routing

The next step opens the backend flow, where the synthesized RTL design undergoes floorplanning, placement, routing, and clock insertion.

Performance and manufacturability verification

Performance and manufacturability verification is the last step of the SoC ASIC design flow. Here, the physical view of the design is extracted. Then, the design undergoes a timing verification process, signal integrity, and design rule checking, which completes the backend design flow.

Summary

In this chapter, we introduced the history behind the FPGA technology and how disruptive it has been to the electronics industry. We looked at the specific hardware features of modern FPGAs, how to choose one for a specific application based on its design architectural needs, and how to select an FPGA based on the Xilinx market offering.

Then, we looked at the history behind using SoCs for FPGAs and how they've evolved in the last two decades. We looked at the MicroBlaze, PowerPC 405, and PowerPC 440-based embedded system offerings from Xilinx and when they switched to using ARM processors in FPGAs. Then, we focused on the Xilinx Zynq-7000 SoC family, which is built around a PS using a Cortex-A9 CPU cluster. We enumerated its main hardware features within the PS and how it is intended to augment them using FPGA logic to perform hardware acceleration, for example. We also looked at the latest generic Xilinx SoC for FPGA and, specifically, the Zynq UltraScale+ MPSoC, which comes with a powerful quad-core Cortex-A53 CPU cluster that's combined in the same PS with a dual-core Cortex-R5F CPU cluster, a flexible interconnect, and a rich set of hardware blocks. This can help provide a good start for many modern and demanding SoC architectures.

Finally, we introduced SoCs for ASICs and how different they are from the SoCs in FPGAs in terms of their design, the associated costs, and the opportunities for each. We also introduced the SoCs in ASICs design flow. Following on from this, in the next chapter, we will introduce the Xilinx SoCs design flow and its associated tools.

Questions

Answer the following questions to test your knowledge of this chapter:

1. Describe the concept upon which the FPGA HW is built.
2. List five of the main hardware features found in modern FPGAs.
3. Which architecture is the Cortex-A9 built on and in which Xilinx FPGA they are integrated?

4. What is the coherency domain that can be defined within the Zynq-7000 SoC FPGA?

5. Describe the coherency domains within the Xilinx UltraScale+ MPSoC FPGA.

6. When would it be suitable to use an FPGA-based SoC for a product?

7. When would an ASIC SoC be a better solution for building a product?

2

FPGA Devices and SoC Design Tools

This chapter will begin by providing an overview of the Xilinx FPGA hardware design flow in general and the tools associated with it, before highlighting the specific tools that are used when designing an SoC for FPGAs. We will also introduce the SoC design hardware verification using the available simulation tools. Finally, we will cover the high-level software design flow and its different steps and introduce the tools involved in the software design for an FPGA-based SoC.

In this chapter, we're going to cover the following main topics:

- FPGA hardware design flow and tools overview

- FPGA SoC hardware design tools

- FPGA and SoC hardware verification flow and associated tools

- FPGA SoC software design flow and associated tools

Technical requirements

The GitHub repo for this title can be found here: `https://github.com/PacktPublishing/Architecting-and-Building-High-Speed-SoCs`.

Code in Action videos for this chapter: `http://bit.ly/3fRWh8z`.

FPGA hardware design flow and tools overview

This section introduces the steps involved in the FPGA hardware design and provides an overview of the associated tools, along with each step.

FPGA hardware design flow

The FPGA hardware design process is similar to the ASIC hardware design process that we covered in the previous chapter. Besides the high ASIC NRE cost and its long time to market path compared to an FPGA design process, the main technical difference is that the choice of features the designer can use and combine to produce a working FPGA device is limited by what the FPGA device itself can offer. However, there is a rich list of devices and features that usually meet many demanding and challenging system requirements. Once the FPGA design specification has been finalized and the target FPGA device has been chosen, the design flow will look as follows:

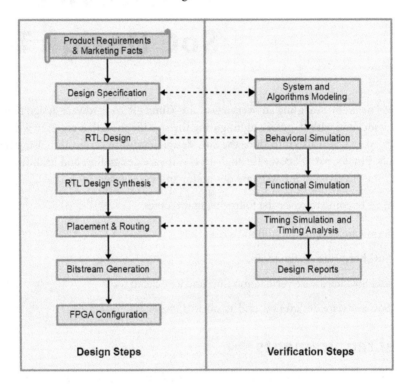

Figure 2.1 – FPGA hardware design flow

Product requirements phase

This is not a design step but rather a fundamental starting phase from which the idea of using an FPGA may have originated. Marketing facts and time-to-market guidelines are studied and a decision on the overall design budget and strategy is put in place to help the engineering teams start working on their parts.

Design specification

This step is also called the **design capture** step, as covered by the *SoC in ASICs design flow* section of the previous chapter. It is the first phase of designing an FPGA and consists of capturing the specification, HW/SW partitioning, IP selection, and defining the external interfaces. The SW portion is usually implemented externally to the FPGA in another ASIC-based SoC or a discrete CPU system. The interfacing mechanisms and the external SW to the FPGA hardware communication stack are defined in detail in this step. Alternatively, if the SW has been implemented within the FPGA, then we are in the FPGA SoC design flow paradigm, which we will cover in the next section. There isn't much of a difference at the architecture level besides the interface's implementation and the fact that the SW is mapped to an ARM CPU or CPUs within the **processing subsystem** (**PS**) portion of the FPGA device. Like the ASIC design flow, the design capture process could be in text format as an architecture specification document. It is also usually associated with a design capture of the specification in a computer language such as C, C++, Python, SystemC, TLM, SystemVerilog, or a combination. Many environments exist where the specification can be expressed in these languages, as follows:

- Imperas OVP, which uses the C language to build and connect system **intellectual property** (**IP**) models. For more information on this environment, please check out `https://www.imperas.com/dev-virtual-platform-development-and-simulation`.

- Synopsys Platform Architect, which uses SystemC and TLM to build and interconnect IP models. For more information on this approach, please check out `https://www.synopsys.com/verification/virtual-prototyping/platform-architect.html`.

- gem5, which uses C++ to build the system IP models and Python to specify a system model and connect the IP models. For more information on this environment, please check out `https://www.gem5.org/documentation/learning_gem5/introduction/`.

The design capture isn't usually a full SoC system model – it could be just an overall description of the main algorithms and inter-block data and command passing mechanisms, which is mainly done in SystemVerilog, to prepare the overall FPGA design verification step.

For the FPGA hardware-centric applications such as those concerned with video and image processing that's performed in the FPGA logic, complex algorithm modeling is performed using third-party tools. The translation tools that produce the RTL are often exploited as-is with the RTL module or a version of it that's been manually tuned for the FPGA technology target device. This technique is also used in software-centric designs seeking to hardware-accelerate software (using the FPGA logic) when discovering bottlenecks while purely executing on a CPU. The latter part of this book will cover both techniques as part of designing advanced systems in FPGA SoCs.

RTL design

Like the ASIC design flow, the design capture phase is followed by the RTL design phase in Verilog, SystemVerilog, or VHDL of the FPGA modules. Their top-level files instantiate and interconnect all the modules. It is worth noting that this step is becoming a team effort due to the increased complexity of what can be implemented in a modern FPGA. Many tools' facilities are available to make it a smooth cooperative effort.

As mentioned previously, some of the RTL is automatically produced by other software tools from a higher-level language, ready to be synthesized as an IP or part of it. High-level synthesis from C, C++, or SystemC is also supported by the tools so that you can directly contribute to generating the IP netlist.

The system design in RTL is a team effort where it is partitioned into multiple blocks. Each block will have an individual owner who coordinates their work with the other members of the team using a design code repository such as Git or Subversion, for example.

RTL behavioral simulation

The RTL design phase is then simulated using test benches written specifically to verify the functional correctness and intended functionality of the RTL design. This simulation usually involves an assertion-based verification methodology. Many RTL simulation tools support the Xilinx FPGA devices, alongside the integrated version of the simulation tools within the Xilinx software design environment. We will cover these tools in detail in a subsequent section of this chapter.

RTL design synthesis

The synthesis tool takes the FPGA RTL design files and generates a netlist, which is a machine file that translates the RTL onto specific elements of the FPGA features that have been mapped using the synthesis software library. The synthesis tool performs a complex translation job by inferring elements from the FPGA features as it recognizes a certain specific characteristic that matches them. In certain situations, it just replaces the RTL with a netlist notation when the RTL style that's used is a direct instantiation of the desired library elements. This technique is sometimes used as an optimization technique for area or speed, where the synthesis algorithm fails to recognize the elements from a high-level behavioral description in the RTL. Hard block macros are also an example of this replacement mechanism at the synthesis stage.

Netlist functional simulation

The functional simulation is usually performed after the design synthesis phase, where the RTL design has been translated into a netlist for grouping the FPGA features. At this stage, it is no longer the RTL behavioral design we are simulating but rather the netlist translation, which has assembled many elements of the FPGA features for implementing the RTL design. It is necessary to perform this step to make sure that the translated outcome matches the intended RTL behavior and that there is no need to direct the synthesis tools to do a better job or optimize in places.

Placement and routing

In this phase, the netlist is distributed across the FPGA features, and elements are geographically placed and connected using the network of routing resources. The **Place and Route (P&R)** software is a complex part of the FPGA technology, and its efficiency is in finding the optimal placements to interconnect the design elements and meet the performance requirements described via the design constraints. This step is usually attained by the software after many attempts and can propose multiple results with different placement and routing. Size, speed, and power are universal metrics that drive the optimization target of the P&R tools. The tools can be requested to use a balanced approach as the design optimization objective.

Given the complexity and size of modern FPGA designs, the P&R stage can be performed on partitions of the design where limited changes have been introduced to the design once its implementation has been started. This won't cause a long tool runtime like that of starting the implementation process afresh. This is called implementation with incremental compile, which can be specified in this situation. It can reduce place and route runtimes and preserve existing implementation optimal results.

Timing simulation

The timing simulation is usually performed after the P&R phase, where the netlist has been mapped to the FPGA resources and the connectivity between the design elements has been established in a scenario that meets the design constraints specified by the designer. At this stage, we are no longer simulating the RTL behavior or its logical representation as a netlist – it is the model of the physical implementation that is being simulated. This accounts for the elements' timing characteristics and response, as well as the routing nets physical delays. These delays introduce all sorts of internal skews and irregularities that we try to visualize in a timing-driven simulation and check that the physical implementation still meets the design timing constraints. Any issues that the tools couldn't resolve are spotted and corrective actions are taken – either manually by manually floor-planning regions of the design or iteratively by going back to the RTL or even to the micro-architecture and resolving the issues at the source.

Timing analysis

This analytical step is performed on the Timing Reports and guided by the Timing Simulation in areas of concern that the designer needs to study to understand the timing characteristics and budget of the design. Areas of concern are resolved, as mentioned in the *Timing simulation* subsection, but sometimes, it is an issue that can be resolved by upgrading the FPGA device's speed grade when the timing budget indicates that the problems or the observations are global within the design.

Bitstream generation

This is the final stage the tool goes through. It generates the binary file that will configure the FPGA to behave as the design desires. There are a few options that we will cover later regarding the method by which the FPGA device will be programmed and its startup sequence characteristics. These are specified as options that the bitstream generation tools must account for. Security and authentication options are also specified at this stage, which we will cover in *Chapter 11, Addressing the Security Aspects of an FPGA-Based SoC*.

FPGA configuration

When powered up, the FPGA device needs programming to behave as specified by the design via the generated bitstream configuration file. Many methods can be used to configure the FPGA device, and each has a specific interface associated with it. For example, you could go over the JTAG interface of the FPGA when still in the debugging phase of the design and perform the initial electronics board bring-up. We will delve into these for a specific device technology such as the Zynq-7000 SoC, although most of the interfaces are common to all FPGAs except for paths over specific hardware features, such as the PS blocks in the SoC FPGAs, which are unavailable in generic non-SoC FPGA devices.

FPGA hardware design tools

The Xilinx FPGA design environment is called **Vivado**. It is a complete design platform that uses **machine learning** (**ML**) algorithms to reduce the FPGA implementation time. It supports all the steps in a Xilinx FPGA design flow, including creating, verifying, implementing, and validating its design. Xilinx provides two versions of the Vivado design suite:

- The ML version, which is a standard and free edition
- The Enterprise edition, which requires purchasing a license for the extra features and the added supported devices

This book will use Vivado ML version 2021.2, which can be downloaded from the Xilinx website for free and does not require a license. You are encouraged to acquire the Vivado ML from the Xilinx website, check the host machine hardware and software requirements, and follow the instructions to install it, as indicated by the Vivado Release Notes, which you can find at `https://www.xilinx.com/support/documentation/sw_manuals/xilinx2021_2/ug973-vivado-release-notes-install-license.pdf`.

The Vivado ML is an **integrated design environment** (**IDE**) that supports two modes of the design flow: **project mode** (**PM**) and **non-project mode** (**NPM**). In the PM, the Vivado IDE manages the design flow process and maintains its database, whereas, in the NPM, it is the user who manages the design sources and executes the design steps by providing the appropriate commands to the specific tool involved at that specific step – for example, synthesis, simulation, and implementation. Every tool has a given binary name by which it is called from the command line, along with the command options that the user can perform on the input files involved at this specific step of the design flow.

In PM, the Vivado design suite uses a project file (`.xpr`) and directory structure to maintain the design source files, store the results of different synthesis and implementation executions, and track the project's status throughout the design flow. Automatically managing the design data, process, and status requires a project infrastructure.

In contrast, npm is for script-based users who do not want Vivado tools to manage their design data or track their design state. The Vivado tools simply read the various source files and compile the design through the entire flow in memory.

In the tutorials associated with this book, we will be using the Vivado ML IDE in project mode. The Vivado ML IDE includes many parts, as follows:

- Vivado IP Integrator
- Vitis High-Level Synthesis
- Vivado Logic Synthesis
- Vivado Simulator
- Vivado Implementation
- Vivado Verification and Debug ILA
- Vivado Device Programmer
- Vivado **Dynamic Function eXchange (DFX)**

Vivado IP Integrator

The Vivado ML Edition IP Integrator is a graphical and TCL-based development flow. It is a device and platform-aware utility that allows you to auto-connect key IP interfaces, generate subsystems, and check design rules. It allows inter-IP system connectivity at the IP interface level, such as AXI or APB, thus reducing the system connectivity time.

Vitis High-Level Synthesis

The Vitis High-Level Synthesis tool enables C++ design capture and its specification to be directly synthesized and target the Xilinx FPGAs. There is no need for the designer to translate it into an RTL language first. This will help in the prototyping stage and even when using it for production when the synthesis results have been judged as optimal.

Logic Synthesis

Vivado includes an RTL synthesis tool that translates the design RTL files into an FPGA netlist ready to be implemented. It supports design descriptions in SystemVerilog, Verilog, and VHDL. The designer also provides synthesis attributes, synthesis command tool options, and **Xilinx Design Constraints (XDCs)** to help optimize this step in the design flow.

Vivado Simulator

Vivado Simulator is a full-featured mixed-language simulator that supports Verilog, SystemVerilog, and VHDL RTL designs. It is an event-driven simulator that supports both behavioral and timing simulations.

Implementation

Vivado Implementation provides the placement and routing software for Xilinx devices. It takes the generated netlist from the Synthesis step and produces the FPGA configuration file. It also generates many design reports to help with analyzing the device logic and routing resources utilization, timing information, power, and other design quality-related metrics.

Verification and Debug

Vivado can help with bringing up and testing the device in various ways. This includes the necessary debug IPs to put into the design, which will then allow the user to connect them to the debug software running on the host machine at runtime. Consequently, this provides system-level runtime visibility into the hardware, which helps control it by setting up trigger events to capture its states and visualize them in the host software interface.

Dynamic Function eXchange

This feature allows you to partially reconfigure a specific portion of the design without shutting down the whole FPGA and having to reconfigure all of it. Some applications may benefit from this option and use a **Time Domain Multiplexing** (**TDM**) approach to reuse portions of the FPGA to implement multiple functions and deploy them on-demand.

FPGA SoC hardware design tools

An SoC design that targets a Xilinx FPGA such as the Zynq-7000 SoC or UltraScale+ MPSoC uses the Vivado IDE and, specifically, the IP Integrator as the SoC design capture tool. The Vivado **Integrated Logic Analyzer** (**ILA**) is used to debug the hardware interactions on the PL side with the PS side and to establish software and hardware co-debug sessions on the system at runtime. Everything else from a design flow perspective is common to a generic FPGA hardware design flow and uses the same tools to synthesize, simulate, verify, implement, and generate the FPGA bitstream file.

In this section, we'll introduce the Vivado IP Integrator and how it can easily be used to create a sample design, including the PS block and an IP from the hardware library catalog, which will be implemented in the PL side of the FPGA. The sample design to use is shown in the following diagram:

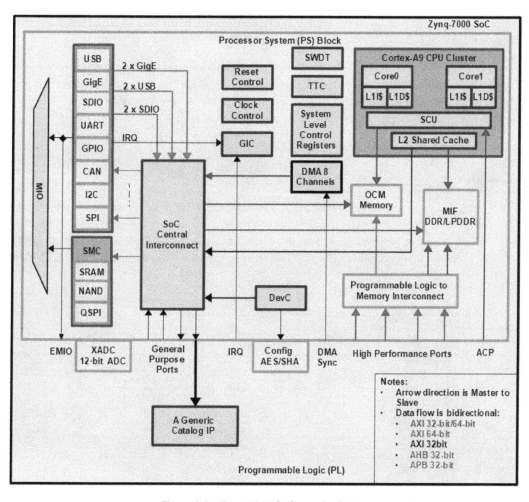

Figure 2.2 – Zynq-7000 SoC sample design

Using the Vivado IP Integrator to create a sample SoC hardware

Let's launch the Vivado IDE and then use its IP Integrator to create the preceding sample design. The intent is to get acquainted with this step of the SoC design capture phase and become familiar with the tools rather than face a challenging design task. You will get the chance to build upon the design's complexity as we progress throughout the different sections of this book and master the fundamental concepts. In this section, you will find steps you can follow to build the sample design. There are three phases involved in this process. First, we must create a design environment, which is simply the Vivado project using the Vivado IDE. Next, we will invoke the IP Integrator to stitch together an SoC, which is formed of the Zynq-7000 PS block and an IP from the IP catalog; the added IP will be implemented in the PL portion of the Zynq-7000. Finally, we must invoke the **Vivado Generate**

Output Products utility to create the HDL files that corresponds to the SoC we have graphically built using the IP Integrator.

For a more detailed step-by-step description of this flow, please refer to the *Vivado Design Suite Tutorial – Embedded Processor Hardware Design*, available at `https://www.xilinx.com/content/dam/xilinx/support/documentation/sw_manuals/xilinx2019_1/ug940-vivado-tutorial-embedded-design.pdf`.

Phase 1 – creating a Vivado IDE project

Follow these steps:

1. First, you must launch the Vivado IDE. Then, on the Vivado launch screen, choose **Quick Start**, then **Create Project**. Give the project a name, such as `SampleProject_2`, specify a location for it in your machine, and click **Next**. Next, specify the **Project Type** option as **RTL Project** and leave everything else as-is, as shown in the following screenshot. Then, click **Next**:

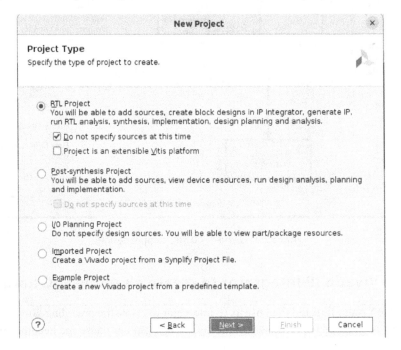

Figure 2.3 – Specifying the Vivado Project Type

2. On the next page, select the **Boards** tab next to the **Parts** tab so that we can target a specific demo board rather than specify everything at the device level ourselves. The catalog of Demo Boards is rich and is also handy for learning purposes as we only need to select the Demo Board that is nearest to our learning objectives. You can order it from its vendor so that you can use it with some of the hands-on chapters in this book. We can also use it as a template and starting

point for our **Proof of Concept (PoC)** or production design. For our exercise, we would like to target a board built around a Zynq-7000 SoC FPGA. As we can see, there are a couple. Here, we will choose **ZYNQ-7 ZC706 Evaluation Board**. Once selected, clicking on **Next** will take us to the **Project Summary** window. The **Project Summary** window gives us an idea of the details that have been captured about the target FPGA SoC device. Now, we need to click on **Finish** so that the project is created. At this point, the project will be created and launched.

Phase 2 – creating a processor subsystem using the Vivado IP Integrator

Follow these steps:

1. On the **Flow Navigator** screen, go to **IP Integrator** and select **[Create Block Design]**. A dialog box will open. Name this IP subsystem design; for example, soc_design_sample_1. Leave the **Directory** field as its default value of <**Local to Project**>. Also, leave the **Specify source set** field as its default value of **Design Sources**. Click **OK** when you're done. The following screenshot illustrates the desired settings:

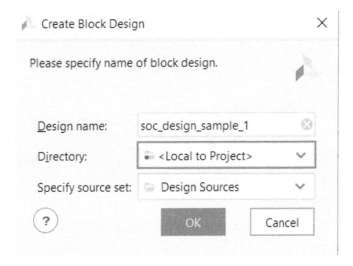

Figure 2.4 – Launching the Vivado IP Integrator

2. In the **Diagram** window, click on the + sign to add an IP to the SoC. Then, using the dialog box that opened, type Zynq in the **Search** field. This will return **ZYNQ7 Processing System** as the only result. Select it by double-clicking on it. This will add the PS to the **Diagram** window. Note the **Designer Assistance** message that is highlighted in the green-colored row on the IP Integrator **Diagram** window with a command hyperlink that states **Run Block Automation**. Click on it. This will open the Design Wizard, which will add the **FIXED_IO** and **DDR SDRAM** interfaces to the Zynq-7000 SoC PS.

3. Now, we can add more IPs manually from the PL side of the FPGA using the IP catalog. Right-click anywhere in the IP Integrator's **Diagram** window and select **Add IP**. In the **Search** field, type bram to filter. You will see that there are two instances: **AXI BRAM** and **LMB BRAM**. Select **AXI BRAM** since we can use this IP to connect to the PS block via one of the generic AXI ports of the PS block. At this stage, the AXI BRAM needs connecting. Click on the **Run Connection Automation** command suggested by the **Designer Assistance** area of the IP Integrator. This will open the IP connection wizard. Select **All Automation**. You will see that the connectivity between IPs is complete. The wizard also added the Reset and AXI SmartConnect IPs.

4. Now, let's configure the size of the BRAM instance and the address mapping of the AXI BRAM controller in the SoC Address Map. Select **Address Editor** in the **IP Integrator** window. We can keep the default values as-is, which will put the AXI BRAM controller at the base address of the general-purpose GPO port facing the PL. The size of the BRAM can be changed to **32KB**.

5. Return to the **Diagram** window. Optionally, you can reshape the system diagram to get a better view by clicking on the **Regenerate Layout** button. Now, let's check that the design capture is fine by running the **Design Rules Check (DRC)**. Go to the Vivado IDE's main menu and click on **Tools | Validate Design**:

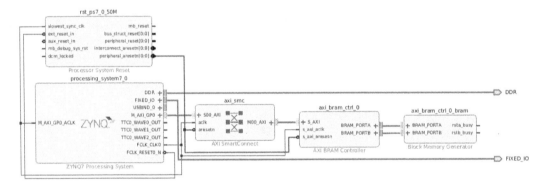

Figure 2.5 – Vivado IP Integrator – the Regenerate Layout button

Phase 3 – generating the HDL files for this design sample

Invoke the **Generate Output Products** utility by right-clicking on the design entry under **Design Sources** in the **Sources** window. Then, click **Generate Output Products**. The design HDL source files will be created.

Now, we need to create the design top-level wrapper. Right-click the top-level subsystem, `soc_design_sample_1`, and select **Create HDL Wrapper** to create a top-level HDL file that instantiates the block design of the sample SoC (PS+IP) we have just created using the IP Integrator tool. This allows us to edit the file if we wish to do so for further integrations, but for our objective, we will choose the default and click **OK**. The top-level wrapper HDL source file will be created and added to the top of the design.

FPGA and SoC hardware verification flow and associated tools

As shown in *Figure 2.1*, the design verification progresses in parallel with the steps of the design flow to make sure that the design's functionality is preserved as the design moves forward in its life cycle. It also ensures that the required **key performance indicators** (**KPIs**) are still met. The KPIs are set at the beginning of the product and system architecture definition. They fundamentally include the system clock frequency, the design's size in terms of FPGA device resources occupation, and the design's energy consumption, among other parameters that the design's performance can be measured by.

The Vivado design environment allows users to perform an RTL behavioral simulation just after the design has been captured in HDL. Once the netlist has been generated, a post-synthesis or functional simulation can be performed and, following the design implementation, a timing simulation can be run. The Vivado IDE provides an integrated simulation tool that can perform all these types of design simulations. Alternatively, the designer can choose to use a third-party HDL simulator for which Vivado IDE can generate run scripts. The Vivado IDE supports all major simulators, as follows:

- Synopsys VCS
- Siemens EDA ModelSim SE/DE/PE, and Questasim
- Cadence Xcelium
- Aldec Active-HDL and Riviera-PRO

In general, to simulate an RTL module of a design, the simulation can be driven by a test bench written in behavioral RTL and instantiating the design to simulate. The test bench is used by the simulator tool to interact with the RTL design by setting its inputs accordingly, thus capturing its outputs at an event-driven execution level. The RTL design is the **design under test** (**DUT**) and the test bench provides input to the actions the simulator will perform during the simulation execution, including the times at which certain tasks are performed. These include, but are not limited to, driving the module's clock input, setting a given signal or input to a specific value at a specific moment in the simulation time for a finite amount of time, and capturing the results of the stimulus on the design.

The Vivado Simulator can also drive the RTL design by forcing its input, capturing its output, and displaying the results as waveforms. There are many options and utilities to choose from in this step of the design verification process. They are mostly suited for pure HDL-based systems than an SoC with a processor and software code to run on them. Nevertheless, this is still a useful verification feature to use when we design an IP and would like to verify it before integrating it into the SoC design. We will cover a practical aspect of the simulation steps and how to benefit from it using a test bench when we study how to integrate a custom IP in the FPGA SoC later in this book.

Adding the cross-triggering debug capability to the FPGA SoC design

This is a useful system verification capability that will allow us to examine the behavior of the hardware that's built into the PL side of the SoC FPGA at runtime as it interacts with the PS side of the SoC. This capability can be added using the ARM CoreSight Debug technology's **Cross Trigger Interface (CTI)** and the **Cross Trigger Matrix (CTM)**, which export the feature outside of the Cortex-A9 CPU cluster.

It is provided in conjunction with the Vivado ILA, which allows us to use the cross-trigger functionality between the Zynq-7000 SoC processor and the hardware IP built into the PL. Cross-triggering allows us to co-debug both the software running on the CPUs within the PS side of the SoC and the soft logic hardware built within the PL side of the FPGA. This means that we can extend the events that can halt the software being executed on the processor when a certain logic combination (as set by the designer and captured at runtime by the ILA IP) is met and vice versa. The following steps illustrate how to add the cross-triggering debug capability to the example SoC design we created previously in this chapter:

1. We must enable **Trigger In** and **Trigger Out** between the PS and PL when we first start creating the design sample and add the PS to the diagram using the Vivado IP Integrator. Now, we can simply remove the PS block from the diagram and add it again while enabling **Trigger In** and **Trigger Out**. The configuration wizard will appear after clicking on the **Run Block Automation** link in the IP Integrator's **Diagram** window:

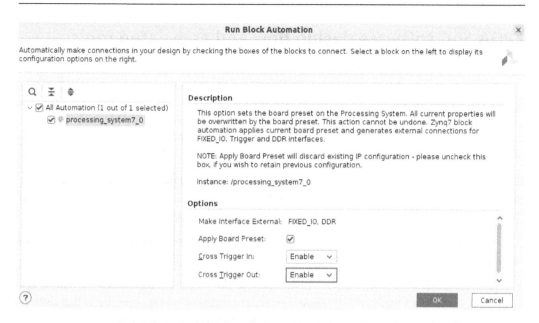

Figure 2.6 – Enabling the PS Trigger In and Trigger Out options

2.   You will notice that **Trigger In** and **Trigger Out** have appeared on the PS interface. Click on the **Run Connection Automation** command, as suggested by the Designer Assistance of the IP Integrator. This will automatically connect the PS back to the SoC with the correct connectivity in place, as shown in the following diagram. With that, we have added the cross-triggering debug feature between the PS and the AXI BRAM controller that's implemented within the PL of the Zynq-7000 FPGA:

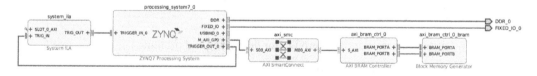

Figure 2.7 – Cross-triggering debug capability

3.   Now, let's generate the HDL for this SoC design, which includes the hardware cross-trigger for runtime debugging. At this point, we can check that the design capture is fine by running the **Design Rules Check** (**DRC**). Go to the Vivado IDE's main menu and click on **Tools | Validate Design**.

FPGA SoC software design flow and associated tools

Software development for an FPGA-based SoC is almost parallel in the SoC's system design to the FPGA SoC's hardware platform design. Once the SoC hardware has been fully captured and verified within Vivado, the software development environment can be handed over. This is another advantage of using an FPGA-based SoC compared to an ASIC where you need to either have an emulation version (and usually a costly one) that targets an FPGA-based prototyping platform or have built a system model for virtual prototyping.

There are handover files that the Vivado IDE creates to use as a base platform for the Xilinx software design environment. This chapter will illustrate all these steps, the files involved, as well as the software utilities used to progress the SoC software development, profiling, and debugging phases.

Xilinx's embedded software development environment is called Vitis. It is the second IDE that centralizes the software building process and manages it as a project.

The Vitis IDE is part of the Vitis unified software platform. The Vitis IDE requires SoC hardware designs built using the Xilinx Vivado IDE. It is based on Eclipse and preserves the familiarity that software developers are acquainted with when building embedded software for discrete microcontrollers. The Vitis IDE's features include the following:

- Project management and source code version control
- C/C++ code editor
- Compilation environment
- Command-line interface
- Application build configuration and automatic Makefile generation
- Error navigation
- An integrated environment with profiles for debugging and profiling
- System-level performance analysis
- Processor Boot Image generation and Flash programming utilities
- FPGA-specific configuration tools

Vitis IDE embedded software design flow overview

The following diagram shows the embedded software application development steps within the Vitis IDE:

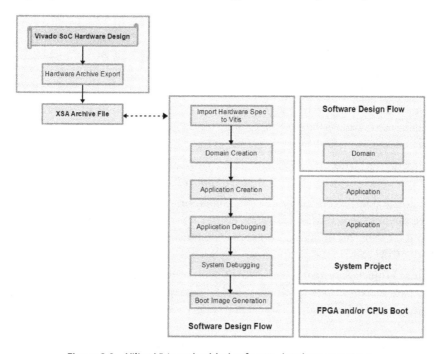

Figure 2.8 – Xilinx Vitis embedded software development steps

Once the hardware design has been completed in the Vivado IDE, the information that's required by the embedded software development is exported from the Vivado IDE to an XSA archive file. The Vitis IDE imports the XSA archive and creates a platform.

The Vitis platform includes the hardware specification and the software environment settings, which are called a domain. The software developer creates applications based on the platform and domains. The software applications can be debugged in the Vitis IDE. Furthermore, in a complex FPGA SoC that encompasses multiple CPU clusters with multiple cores, several applications may run concurrently while employing some type of **inter-process communication** (**IPC**), such as requiring SoC system-level verification. Once all these steps are satisfactory and the SoC FPGA bitstream has been produced within the Vivado environment, the Vitis IDE is used to prepare the boot images that initialize the system and launch the built software applications.

Vitis IDE embedded software design terminology

Before delving into the embedded software design in the following chapters, it is a good idea to become familiar with the main terminology used in this book and across the Xilinx literature related to the FPGA SoC embedded software development flow:

- **Workspace**: This is the Eclipse IDE framework. When the Vitis environment is launched for the first time, a workspace is created. It is simply a directory that's used by the Vitis software platform to store project data and metadata.

- **XSA**: This is the hardware design archive that's exported from the Vivado Design Suite. It contains hardware specifications such as the processor configuration properties, the peripheral connection information, the system address map, and the device initialization code.

- **Platform**: The platform is a combination of hardware components and software components. Software components include the **Board Support Package** (**BSP**) and the boot components.

- **Domain**: A domain combines a BSP or the **operating system** (**OS**) with a collection of software drivers. A domain is a base for building the user application. We can create multiple applications to run on a specific domain. A domain is tied to a single processor or a cluster of multiple similar CPU cores in the platform.

- **System Project**: A system project groups applications that run simultaneously on a device. Two bare-metal software applications for the same processor cannot both reside in a system project, whereas two Linux software applications can.

Vitis IDE embedded software design steps

As illustrated in *Figure 2.8*, first, the hardware design is created with the Vivado IDE. Then, it is exported via the XSA archive file to be used by the Vitis IDE. The software to be run on the FPGA SoC is created step by step, as follows:

1. The platform project is created using the XSA file.
2. The domain is created and added to a platform project.
3. The domain is configured.
4. The application is created.
5. The application is built (compiled).
6. The application is downloaded to the target board (or virtual prototype).
7. The application is debugged and profiled.

We will cover the preceding steps on many of the SoC hardware platforms we will be building in the upcoming chapters.

Summary

In this chapter, we introduced the FPGA and SoC hardware design flow, from defining the architecture and capturing it to generating the FPGA device configuration file. We also looked at the hardware design verification that's involved at every step of the design flow and explained its purpose and how it can be performed. Then, we looked at the SoC design capture in the Vivado IDE and how easily a PS SoC can be created, and how it can be extended using off-the-shelf IPs from the Xilinx IP catalog. We also looked at how hardware and software co-debugging capabilities can be added to the design using the ARM CTI and Xilinx ILA features. We also introduced the SoC software design framework and the Vitis IDE and how a software project can be created using the XSA archive file. Finally, we explored the software design steps and the Xilinx terminology that's used for the FPGA SoC-embedded software development.

The next chapter will address more of the SoC design and architecture fundamentals, such as the major on-chip interconnect protocols used in modern SoCs, the on-chip data movement, which uses DMA engines, and the data-sharing challenges.

Questions

Answer the following questions to test your knowledge of this chapter:

1. List and describe the main steps in the FPGA SoC hardware design flow.
2. List and describe the main steps in the FPGA SoC hardware design verification process.
3. What are the main elements and features that are used to enable SoC hardware and software co-debugging?
4. How important is the co-debugging capability for modern deeply integrated SoCs?
5. How many phases are involved in an FPGA SoC design capture? List them and provide a summary of each phase.
6. Describe the FPGA SoC software design flow.
7. Which design environment is used for designing the software of a Xilinx FPGA-based SoC?
8. List the main steps in a Xilinx FPGA SoC-embedded software development flow.

Basic and Advanced On-Chip Busses and Interconnects

This chapter will begin by providing an overview of the buses and interconnects that are used within an SoC. We will introduce the concepts of data sharing and coherency and how to solve their associated challenges. This will give you a good introduction to the **Advance Microcontroller Bus Architecture (AMBA)** and the **Open Core Protocol (OCP)** protocols. Finally, we will cover the data movement within an SoC and how to use DMA engines.

In this chapter, we're going to cover the following main topics:

- On-chip buses and interconnects overview
- ARM AMBA interconnect protocols suite
- OCP interconnect protocol
- DMA engines and data movements
- Data sharing and coherency challenges

On-chip buses and interconnects overview

FPGA and ASIC-based SoCs are built using multiple components, which are a combination of modules provided as macros by the FPGA vendor, designed in-house usually in RTL, and third-party modules that require a form of licensing to use. These modules are commonly referred to as **intellectual properties (IPs)**. These IPs are connected in a topology specified by the SoC hardware architecture using buses and interconnects. They collaborate, which means they need to interact at runtime to implement a specific set of tasks as part of the system's overall functionality. There are many levels of functional complexity and features that a given interconnect supports. These are based on a bus protocol specification such as ARM AMBA, OCP, or IBM CoreConnect, to mention a few. In this chapter, we will focus on ARM AMBA, which is a collection of bus protocols grouped under a specific AMBA standard revision. However, we will also cover the OCP bus protocol as it is also common in the ASIC SoC designs and you may need to design OCP to AMBA bridges to integrate third-party IPs that use interfaces based on the OCP protocol specification. You may also be porting an ASIC-based SoC to an FPGA-based SoC for prototyping and/or production. This will also help us compare it to ARM AMBA in terms of functional features and complexity.

On-chip bus overview

The communication bus is the medium over which two interfaces from different IPs communicate. A bus can be as simple as a point-to-point single transaction or a single-threaded connection lane or implemented using a complex many-to-many, multi-threaded, and multi-layer transactions protocol. At one end of the bus, there is a transaction initiator or a master interface while at the other end, there is a target or a slave interface. The initiator puts a data access request on the bus that the target responds to by consuming the write data provided by the initiator or by providing the read data requested by the initiator. A simple bus is usually composed of the following lanes:

- A data lane (or lanes for cases where the read path is different than the write path).

- An address lane to specify the exact storage location of the data in the completer or target's address space.

- A control lane to qualify the transaction as a read or write and provide synchronization information as to when a given control signal is valid, ready, and so on.

The following diagram illustrates the concept of a simple bus:

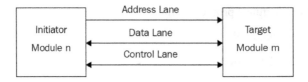

Figure 3.1 – Simple point-to-point bus

On-chip interconnects

Interconnects add many functional features to the point-to-point bus, which make them able to connect many buses in a single shared system address space. This usually requires a central switching capability, bus protocol conversion, buffering, and arbitration agent that can dynamically allow a requesting initiator and a target to establish a connection and exchange data, even when they don't belong to the same bus protocol standard. Interconnect switches differ in complexity from a simple crossbar switch to a **Network-on-Chip (NoC)** with protocol layers conversion, data coherency, advanced routing capabilities, multi-transactions, and multi-threaded capabilities. The following diagram illustrates the concept of a crossbar switch being used to interconnect multiple IPs using simple buses. The simple features are just like those in a point-to-point simple bus case:

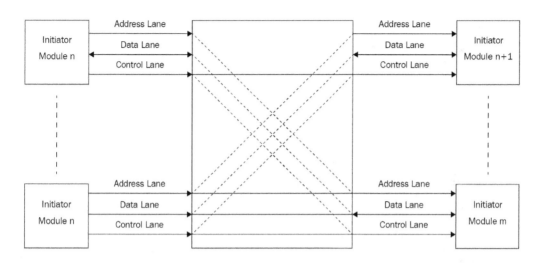

Figure 3.2 – Simple switching matrix-based interconnect

In an NoC, the initiator can be multi-threaded, meaning that multiple hardware threads can share the same bus interface in the NoC. Here, the interconnect can use techniques to track many parallel transactions originating from the same hardware bus interface toward targets, even though they originate from different hardware threads. Every bus protocol that supports this feature has mechanics that allow the NoC, the initiators, and the targets to share the bus medium and fulfill these types of transactions. There is also support for concurrent transactions for both read and write since the bus uses multiple paths for the request phase, the command response phase, and the data phases. In some bus protocols, such as AMBA AXI, there are multiple data paths – one for the read and one for the write, allowing concurrent read and write transactions. It can also issue many of these transactions before waiting for the ongoing ones to complete first, though this is quantitatively limited by the transaction type issuing capability of the initiator, the NoC, and the target. The read and write transactions are also varied and aligned with many kinds of masters, such as DMA engines, processors, and processor caches. The following diagram illustrates the concept of NoC:

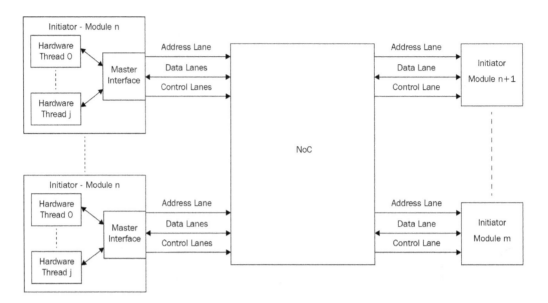

Figure 3.3 – Network-on-chip-based interconnect

ARM AMBA interconnect protocols suite

AMBA is a collection of bus and interconnect specifications provided by ARM for use in SoCs to attach initiators to targets with different levels of features and protocol complexity. It is an open standard and free to use in SoCs. The specification can be accessed from the ARM website at `https://developer. arm.com/architectures/system-architectures/amba/specifications`.

The current revision is revision 5. Historically, each newer revision kept backward compatibility with the previous ones but also added new bus protocols with newer features that were developed to keep pace with the modern SoC complexities and higher performance demands. AMBA had to evolve to support multi-core and multi-cluster CPU topologies, which also use some sort of accelerator with all the challenges they present while maintaining the support for lower performance and simpler SoC topologies. The more complex a hardware system becomes. the more power-hungry it will become. ARM has been providing features that balance all the system **Key Performance Indicators** (**KPIs**) to still meet the needs of building complex SoCs.

ARM AMBA standard historical overview

The first AMBA standard or AMBA specification revision 1.0 (AMBA1) was available from ARM in 1996. It included two bus protocols: the **Advanced System Bus** (**ASB**) and the **Advanced Peripheral Bus** (**APB**). Only the APB bus protocol is still in use in today's SoCs and mostly for registers access and as a configuration interface for IPs. The second revision of the AMBA specification (AMBA2) was released in 1999 and added the **Advance High-performance Bus (AHB)** to the existing ASB and APB bus protocols from revision 1.0. Revision 3.0 (AMBA3) was released in 2003 and added the **ARM eXtensible Interface (AXI-3)** for high-performance data exchanges and **ARM Trace Bus (ATB)** for trace data encoding and transfer capabilities between elements of ARM CoreSight (ARM's SoC debug and tracing technology). AMBA revision 4.0 (AMBA4) was first released in 2010 by launching the AXI-4 bus protocol as an upgrade to AXI-3. In 2011, ARM added coherency support, which resulted in the introduction of the **AMBA AXI-4 Coherency Extension (ACE)** bus protocol. AMBA4 also defines the AXI Streaming protocol, which is used in point-to-point connectivity between IPs for high data transfers. It also added the low power interface specification for clock management and power control. These were known as the *Q-Channel* and *P-Channel* interfaces, respectively. The latest AMBA revision is 5.0 (AMBA5), which was released in 2013. It defines the **Coherent Hub Interface (CHI)** protocol as an interface with the capability to sustain high-performance data exchanges while also interconnecting processors in a cache-coherent way. AMBA5 also upgraded the AXI bus to AXI-5 and, over time, added support for many newer bus protocols, such as the **Credited eXtensible Stream (CXS)** protocol for point-to-point data exchanges between IPs. AMBA5 introduced the **AMBA Adaptive Traffic Profile (ATP)** specification, which isn't a bus interface by itself but rather a qualitative specification associated mainly with the AXI master interfaces in real-time transactions and their timing. AMBA5 includes the **AMBA Generic Flash Bus (GFB)** specification, which is specifically designed to support non-volatile memories such as flash device transaction types. AMBA5 also includes the **AMBA Distributed Translation Interface (DTI)**, which defines the protocol that's used by elements of the system memory management units for system-level address translation services within the SoC. The following diagram visualizes the historical evolution of the AMBA standard and summarizes the specific bus and interconnect protocol within each revision of the standard:

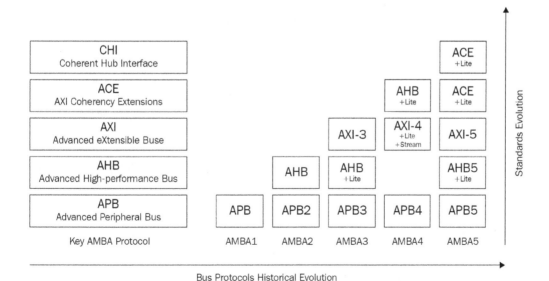

Figure 3.4 – AMBA interconnect standards and bus protocol evolution

APB bus protocol overview

This section will explore the APB bus protocol, its evolution throughout the different AMBA standard revisions, and its added features and mechanisms. We will gain an understanding of this bus's supported transactions, signaling, and application use cases. We will provide an example system implementation using the APB bus.

APB bus protocol evolution

The APB bus is the simplest interface protocol included in the ARM AMBA standard. It has evolved since its first inception in 1996. Every revision added new features and improvements to the protocol while keeping backward compatibility with the previous revisions. APB revision 1.0 is now obsolete and only APB2, APB3, APB4, and APB5 are still active protocols in the industry.

APB2 is considered the base APB protocol for defining the signal interfaces, the read and write supported transactions, and the two APB available components, namely the APB bridge and the APB slave.

The APB3 protocol added support for wait states and transaction error reporting to the base APB2 protocol. The PREADY and PSLVERR signals were added to allow this protocol to expand.

In the APB4 protocol, transaction protection and sparse data transfer features were added. The PPROT signal implements the secure transaction access, so it is used to distinguish between secure and non-secure transactions over the APB bus, while the PSTRB signal implements the write strobe to enable the sparse data write transaction between an AMBA master and an OPB slave.

The APB5 revision extends the APB4 protocol with support for wakeup signaling, user signaling, parity protection, and check signals. These features can enhance the SoC's power consumption, extend the APB protocol by using custom sideband signaling, and improve reliability in the system.

APB bus characteristics

The APB protocol is a simple interface that requires minimal silicon resources for its implementation in an SoC compared to other AMBA bus protocols, thus making it a low-power bus. Data transfers on the APB bus require a minimum of two clock cycles. The APB bus is designed to be a side or secondary bus through which the CPU can implement a control path, thus avoiding any interference from the main data bus, which is implemented using one of the high performances AMBA buses such as AXI or AHB. At runtime, the CPU can use the APB bus to set up an IP registers file, read the status of IP transactions that have been completed, or any other control path-related tasks that software can easily split from the data path to free it for high throughput access, such as I/O packet data or memory data structures through the CPU caches. The IP APB ports are usually grouped as a tree of up to 16 ports that hang off an APB bridge from the main SoC interconnect. The APB bridge performs the transactions protocol conversion and mapping. The APB bridge acts as the transaction's initiator in this topology, while the APB slave behaves as the target that responds to these transactions. The APB specification also refers to the APB bridge as the Requester and the APB slave or target as the Completer.

APB bus interface signals

The following diagram illustrates the connectivity between an APB Requester and an APB Completer, where all the signals defined by APB5 are present:

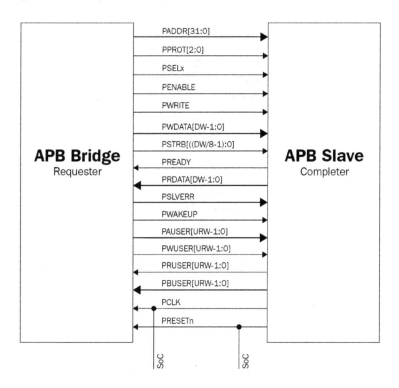

Figure 3.5 – APB bus interface signals

As shown in the preceding diagram, the APB bus signals are split into three categories: the Requester-driven data lane and signals, the Completer-driven data lane and signals, and the SoC bus clock and reset signals. The following table provides a short description of the Requester signals:

Signal	Width	Description
PADDR	32	Address Lane, driven by the Requester.
PPROT	3	Protection type control signal. It indicates if this is a normal, privileged, or secure operation. It is also used to indicate if this is data or instruction access.
PSELx	1	Select signal. Indicates to the Completer that it has been selected by the Requester to respond to an incoming transaction.

Signal	Width	Description
PENABLE	1	Enable signal. It is set to indicate the start of the second cycle in an APB transaction.
PWRITE	1	Transaction direction. When HIGH, it is a Write; when LOW, it is a Read.
PWDATA	DW	Write the data lane of DW width. The Requester uses this lane to place the write data that targets the Completer. DW is usually 8, 16, or 32 bits.
PSTRB	DW/8	Write strobe signals. Each signal indicates that the corresponding data byte is valid.
PWAKEUP	1	Wake-up signal. This signal is used by the Requester to indicate activities associated with this APB interface.
PAUSER	URW	User-defined request attribute lane. This lane can be used to extend the APB protocol and customize it to support other features and attributes not defined by the APB5 protocol. This lane represents the request phase of a custom-defined transaction.
PWUSER	UDW	User-defined write attribute lane. This lane can be used to extend the APB protocol and customize it to support other features and attributes not defined by the APB5 protocol. This lane can carry the write data from the Requester to the Completer of an APB custom-defined write transaction.

Table 3.1 – APB bus interface signals description

The following table provides a short description of the Completer signals:

Signal	Width	Description
PREADY	1	Ready signal. This signal is used by the Completer to qualify or extend the APB transaction on the bus.
PRDATA	DW	Read data lane. Used by the addressed Completer to provide the requested read data. It is usually 8, 16, or 32 bits wide.
PSLVERR	1	Transfer error signal. Used by the slave to indicate the transaction completion status to the Requester.
PRUSER	UDW	User-defined read attribute lane. This lane can be used to extend the APB protocol and customize it to support other features and attributes not defined by the APB5 protocol. This lane can carry the read data returned by the Completer to the Requester of an APB custom-defined read transaction.

Signal	Width	Description
PBUSER	URW	User-defined response attribute lane. This lane can be used to extend the APB protocol and customize it to support other features and attributes not defined by the APB5 protocol. This lane represents the response phase of a custom-defined transaction.

Table 3.2 – APB bus interface Completer signals description

The following table lists the SoC signals:

Signal	Width	Description
PCLK	1	Clock signal. This is a common signal provided by the SoC to both the Requester and Completers.
PRESETn	1	Reset signal. This is a common active LOW signal provided by the SoC to both the Requester and Completers.

Table 3.3 – APB bus interface SoC signals description

APB bus-supported transactions

The APB bus's latest protocol – that is, APB5 – defines the following transaction types:

- Write transactions without the *Wait* state
- Write transactions with *Wait* state insertion
- Read transactions without the *Wait* state
- Read transactions with *Wait* state insertion
- Write transactions with *Write* strobes
- Error response
- Secure and non-secure transactions
- Wake-up transaction signaling
- User signaling

This section only attempts to provide a functional overview of the preceding supported transactions in APB5. You are encouraged to study the AMBA5 standard that defines the APB5 bus protocol for further implementation details. The APB5 bus protocol can be found at https://developer.arm.com/documentation/ihi0024/latest/.

Write and read transactions are simple back-to-back transactions that are set by the Requester in a predefined manner so that the Completer can sense them and decode the transaction accordingly. Data is then accepted from the write lane or provided on the read lane to the Requester. As mentioned previously, the APB bus is a simple transaction bus with no pipelining or burst support. The protection support is worth highlighting as it adds an orthogonal qualification to the transaction using the PPROT bits, as follows:

- PPROT[0]: This indicates if this is a Normal or Privileged transaction and usually reflects the status of the running state of the Requester. It can easily be mapped by the CPU to its execution state (kernel mode or user mode, for example).

- PPROT[1]: Secure or non-secure. This signal can be used to implement hardware security by allowing access to certain registers when the transaction is classified as secure – that is, PPROT[1] is LOW. When PPROT[1] is set to HIGH by the Requester, access to secure registers, for example, is prohibited by the Completer.

- PPROT[2]: Data or instruction. This signal provides a hint regarding the type of data exchanged by the transaction.

APB bus example system implementation

As mentioned previously in this section, the APB Requester is an APB bridge that translates an upstream protocol, such as AHB or AXI, into an APB bus protocol to provide a control path to a CPU through which it can set up IP peripheral registers and check the runtime status following an interrupt event, for example. The following diagram provides an overview of such an implementation, where the APB bus is connected to the SoC interconnect through an AXI to APB bridge:

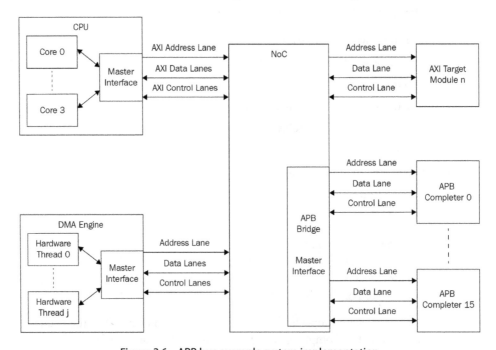

Figure 3.6 – APB bus example system implementation

AXI bus protocol overview

This section will explore the AXI bus protocol, its evolution throughout the different AMBA standard revisions, and its added features and mechanisms. We will gain an understanding of this bus's supported transactions, signaling, and application use cases. We will provide an example system implementation using the AXI bus. This section tries to provide a detailed enough overview of the AXI-3 and AXI-4 bus protocols as these are used in the Zynq-7000 SoC and Zynq UltraScale+ SoC FPGAs. AXI-5 is mentioned for completeness only, but you are still encouraged to study it using the previously mentioned AMBA5 standard specification from ARM.

AXI bus protocol evolution

As illustrated in *Figure 3.4*, the AXI bus protocol first appeared in the AMBA3 standard as a high-performance bus interface to become the ARM-based SoC main data path interconnect.

AMBA4 added more features and supporting channels and signals to AXI to produce AXI-4, a richer bus protocol with backward compatibility and interoperability with AXI-3. AXI-4 extended the burst lengths from 16 to 256 beats, removed the write interleaving and the locked transaction support features introduced in AXI-3, and added the notion of transaction **Quality of Service (QoS)** support. Like in APB5, AXI-4 added support for user-defined channels using the AxUSER (address), WUSER (write data), RUSER (read data), and BUSER (response channel) lanes. These can be used to extend the AXI-4 protocol and customize it to support other features and attributes not defined by the base AXI-4 protocol. AXI-4 also added support for regions using the AWREGION and ARRREGION vectors, which can encode up to 16 regions. Consequently, the same AXI-4 slave port can implement multiple logical interfaces mapped to a different region in the system address map.

AMBA4 also defined the AXI Streaming protocol, which allows high throughput point-to-point data exchanges between two connected IPs.

AMBA4 also added AXI Lite, which is a simplified version of the full AXI bus protocol.

AMBA5 defined AXI-5, which, in the latest AXI revision, provides even higher performance, scalability, and a wider feature choice for system IPs data and signaling exchanges in an SoC when using AXI as an interconnect between masters and slaves. AXI-5 extends support for atomic transactions, cache stashing and de-allocation, and data protection and poisoning signaling. It also added support for persistent **cache maintenance operations (CMOs)**, among many other features and options.

AXI bus characteristics

The AMBA AXI protocol is a high-performance and high-speed interconnect for modern SoC designs. It provides a high-throughput and low-latency system bus for highly demanding CPU clusters. It can be easily interconnected via simple bridges in the SoC interconnect with existing AHB and APB-based IPs. The AXI interface relies on multiple lanes with separate address/control and data phases to initiate a transaction. It supports unaligned data transfers by using byte strobes. AXI transactions use bursts to transfer data, which only requires a single start address. AXI is a true full-duplex bus that uses two separate channels – one for write transactions and one for read transactions. They also support multiple concurrent transactions to different addresses with an out-of-order completion capability.

AXI bus interface signals

The AXI bus uses five channels, as follows:

- The write address
- The write data
- The write response
- The read address
- The read data channels

It also has a low-power interface. There are also global signals, namely clock and reset, which are driven by the SoC. The following diagram depicts the AXI bus multi-channel topology:

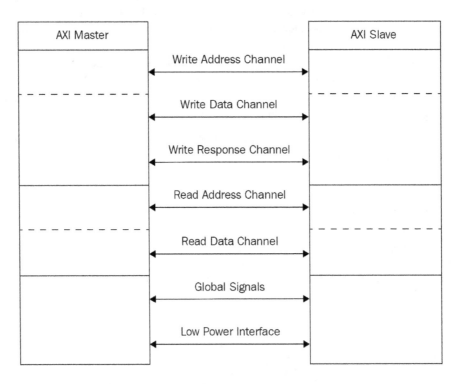

Figure 3.7 – AXI bus multi-channel topology

AXI bus global signals

The following diagram illustrates the connectivity of the global signals:

Figure 3.8 – AXI bus global signals connectivity

AXI write address channel signals

The following diagram illustrates the connectivity of the AXI write address channel signals:

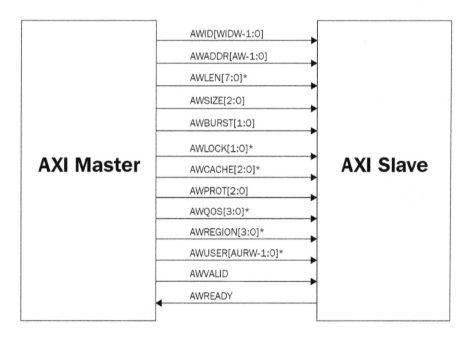

Figure 3.9 – AXI bus write address channel signals connectivity

Signals with an asterisk (*) next to them mean that they present differences between AXI-3 and AXI-4.

AXI write data channel signals

The following diagram illustrates the connectivity of the AXI write data channel signals:

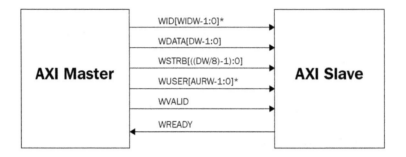

Figure 3.10 – AXI bus write data channel signals connectivity

Signals with an asterisk (*) next to them mean that they present differences between AXI-3 and AXI-4.

AXI write response channel signals

The following diagram illustrates the connectivity of the AXI write response channel signals:

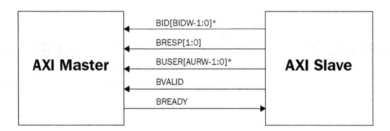

Figure 3.11 – AXI bus write response channel signals connectivity

Signals with an asterisk (*) next to them mean that they present differences between AXI-3 and AXI-4.

AXI read address channel signals

The following diagram illustrates the connectivity of the AXI read address channel signals:

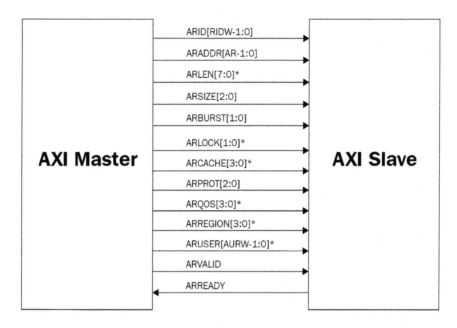

Figure 3.12 – AXI read address channel signals connectivity

Signals with an asterisk (*) next to them mean that they present differences between AXI-3 and AXI-4.

AXI read data channel signals

The following diagram illustrates the connectivity of the AXI read data channel signals:

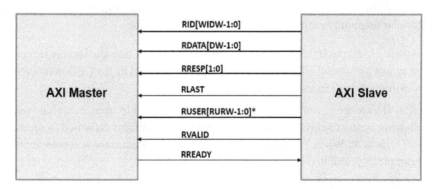

Figure 3.13 – AXI bus read data channel signals connectivity

Signals with an asterisk (*) next to them mean that they present differences between AXI-3 and AXI-4.

AXI low-power interface signals

The following diagram illustrates the connectivity of the AXI low power interface signals:

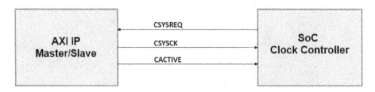

Figure 3.14 – AXI low-power interface signals connectivity

AXI bus-supported transactions

An AXI transaction is initiated by an AXI master toward an AXI slave performing either a read operation, a write operation, or a user-defined operation. The payload data is moved over the AXI bus using an AXI burst, which is formed by AXI beats. Three types of burst transfers are supported: *FIXED (0b00)*, *INCR (0b01)*, and *WRAP (0b10)*, as specified by the *AxBURST[1:0]* vector (*x* being *R* for the *read* transactions and *W* for the *write* transactions). *AxLEN[7:0]* (*x* being *R* for *read* and *W* for *write*) specifies the burst length, which can be 1 to 16 transfers in AXI-3 for all burst types. It is 1 to 16 transfers in AXI-4 for the FIXED and WRAP bursts, though the INCR burst can be from 1 to 256 transfers for AXI-4. In addition to the size and layout, some attributes are associated with transactions, as defined by the *AxCACHE[3:0]* vector (*x* being *R* for *read* and *W* for *write*). This vector helps in qualifying the memory and peripheral slave support and, as such, how the system handles them while progressing through interconnects and bridges. It also specifies how the system-level caches deal with them when involved.

The AXI protocol also supports protection via the *AxPROT[2:0]* vector (*x* being *R* for *read* and *W* for *write*) in a similar mapping to the *PPROT[2:0]* in the APB protocol:

- *AxPROT[0]*: Indicates if this is a normal or privileged transaction and usually reflects the status of the running state of the Requester. It can easily be mapped by the CPU to its execution state (kernel mode or user mode, for example).

- *AxPROT[1]*: secure or non-secure. This signal can be used to implement hardware security by allowing access to certain registers when the transaction is classified as secure – that is, *PPROT[1]* is LOW. When *PPROT[1]* is set to HIGH by the Requester, access to secure registers, for example, is prohibited by the Completer.

- *AxPROT[2]*: Data or instruction. This signal provides a hint regarding the type of data exchanged by the transaction.

The AXI protocol adds the possibility of transaction processing and progressing concurrency if they are not tagged with the same AXI ID. All transactions from the same master that have been tagged with the same AXI ID must be processed by the slave and sent back to the issuing master in the same order they first arrived. However, transactions from the same master toward the same slave and with different AXI IDs can be processed and completed out of their arriving order. This property allows for better system performance if progress can be made outside of the processing ordering rule.

> **More Information**
>
> For a detailed description of the AXI transaction characteristics, you are encouraged to study the AXI-4 and AXI-3 protocol specifications, which are available at https://developer. arm.com/documentation/ihi0022/hc/?lang=en.

AXI bus example system implementation

A good example of using AXI-3 as the main bus protocol of an SoC can be seen in the following Zynq-7000 SoC interconnect diagram. The control path uses the APB bus to control peripheral registers and check the IP status at runtime. Memory and high throughput peripheral paths are implemented using the AXI-3 bus and its interconnects:

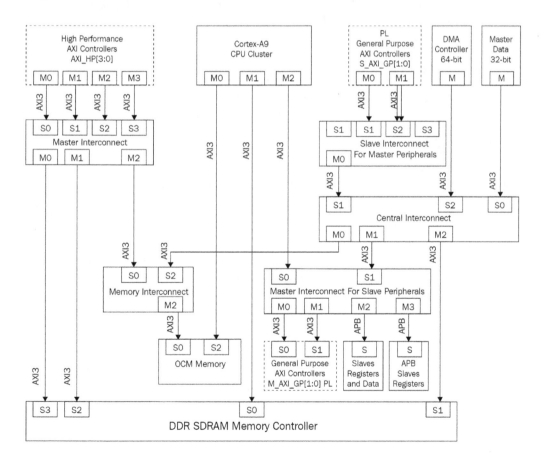

Figure 3.15 – AXI interconnect-based system example – the Zynq-7000 SoC

AXI Stream bus protocol overview

The AXI Stream bus protocol was first introduced by ARM in AMBA4. It is a serial point-to-point link that connects two IPs for data and control information exchange. This section will explore the AXI Stream bus protocol, its evolution from AMBA4 to AMBA5 standards, and its added features and mechanisms. We will understand this bus's supported transactions, signaling, and application use cases. We will provide an example system implementation using the AXI Stream bus.

AXI Stream bus protocol evolution

At the time of writing, AXI Stream in the AMBA5 standard is at its second revision. The first revision appeared in AMBA4 as a way to connect two IPs to exchange data in a single direction – that is, one interface acting as a master and the other interface acting as a slave. AXI Stream in AMBA5 added two main features: wake-up signaling and parity protection. To build bidirectional data exchange mechanisms between two IPs, two AXI Stream buses can be used by deploying one in each direction. An AXI Stream bus can cross an interconnect, but often only for data width adaptation, such as upsizing or downsizing. It can also perform clock domain crossing when the master and the slave clocks belong to two different clock domains.

AXI Stream bus characteristics

The AXI Stream protocol defines three types of bytes: a *data byte*, which is the information to be transferred from a master interface to a slave interface, a *position byte*, which indicates the relative positions of data bytes in a stream of data, and a *null byte*, which is not a data nor a position byte. These bytes are qualified by the TKEEP and TSTRB signals on the AXI Stream bus, as indicated by *Table 3.4*, where the slave can discard null bytes and use position bytes to format the data transfer sequences or implement a higher custom protocol transparently on the underlying AXI Stream bus protocol.

AXI Stream also defines a *transfer* to exchange data through the AXI Stream interface. This is done within a *TVALID* and *TREADY* handshake between the master and slave interfaces. A *packet* contains several data bytes that belong to the same transfer or transfers, similar to a burst transaction in the AXI-4 bus protocol. A *frame* is a logical grouping of multiple packets with each assembling several data bytes. A *data stream* is used to transfer data from a master to a slave as many individual byte transfers or as a grouping of packets. The AXI Stream protocol supports three types of handshake sequences between the master and the slave:

- Handshake with TVALID, as asserted by the master before TREADY is asserted by the slave.
- Handshake with TVALID, as asserted by the master after TREADY is asserted by the slave.
- Handshake with TVALID and TREADY, asserted concurrently.

AXI Stream bus interface signals

This section summarizes the signals that form the AXI Stream bus interface for both the master and the slave. Taking a closer look at these signals will provide a good overview of the protocol mechanics to move data from source to destination. However, you are encouraged to examine the protocol specification from ARM at `https://developer.arm.com/documentation/ihi0051/b/?lang=en`.

The following table lists the AXI Stream bus interface signals:

Signal	Width	Description
ACLK	1	Interface clock signal that's used for sampling all the bus signals on the rising edge.
ARESETn	1	Interface global reset, active low.
TVALID	1	A master interface signal indicating that data is valid on the TDATA lane.
TREADY	1	A slave interface signal indicating that the slave is ready to accept the data on the TDATA lane.
TDATA	TDW	The bus data lane's size, which can be 8, 16, 32, 64, 128, 256, 512, or 1,024 bits.
TSTRB	DW/8	A byte qualifier indicating that the associated byte in the TDATA lane is a data byte or position byte type.
TKEEP	DW/8	A byte qualifier indicating that the associated byte in the TDATA lane is a data/position byte or a null type. This indicates if the associated byte needs processing at all.
TLAST	1	Packet delimiter.
TID	TIDW	Used as a data stream ID with TIDW <= 8.
TDEST	TDESW	Helps route the data stream through AXI Stream topologies that have multiple hopes in their paths. TDESW <= 8.
TUSER	TUW	Sideband information that's used to extend the AXI Stream protocol by the user.
TWAKEUP	1	Part of the AXI-5 Stream protocol that's used to indicate that there is activity in the AXI-5 Stream bus.

Table 3.4 – AXI Stream bus interface signals description

AXI Stream bus-supported transactions

The AXI Stream protocol defines four styles of data streams: the *byte stream*, the *continuous aligned stream*, the *continuous unaligned stream*, and the *sparse stream*. Let's look at each in detail.

Byte stream style

In a byte stream transfer style, all the byte types can be used. The following diagram illustrates a byte-stream transfer, where the interface is 4 bytes wide:

Progress of time

	Word 0	Word 1	Word 2	Word 3	Word 4	Word 5	Word 6
	NULL	NULL	B-06	B-10	NULL	B-16	B-19
	NULL	B-03	NULL	B-09	B-13	NULL	NULL
	B01	NULL	B-05	B-08	B-12	B-15	B-18
First transferred byte	B00	B-02	B-04	B-07	B-11	B-14	B-17

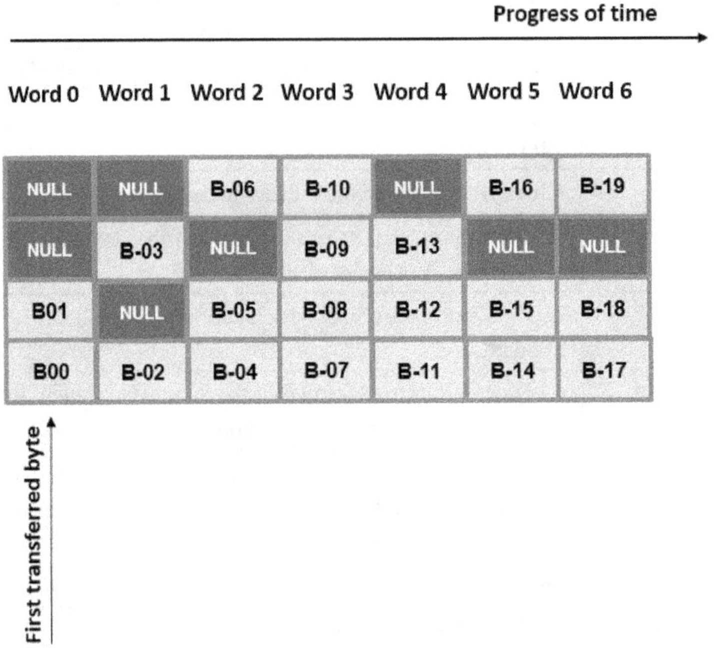

Figure 3.16 – AXI Stream protocol byte stream example

In the preceding diagram, *Word 0* is the first four bytes to be transferred across the bus from the master to the slave and it has *B00*, *B01*, and two null bytes forming it. This is followed by *Word 1*, *Word 2*, and so on.

Continuous aligned stream style

In a continuous aligned stream, only the data byte type can be used. The following diagram shows a continuous aligned stream transfer, where the interface is 4 bytes wide:

	Word 0	Word 1	Word 2	Word 3	Word 4	Word 5	Word 6
	B03	B07	B11	B15	B19	B23	B27
	B02	B06	B10	B14	B18	B22	B26
	B01	B05	B09	B13	B17	B21	B25
	B00	B04	B08	B12	B16	B20	B24

Passage of time

Figure 3.17 – AXI Stream protocol continuous aligned stream style example

In the preceding diagram, *Word 0* is the first four bytes to be transferred across the bus from the master to the slave and it has *B00*, *B01*, *B02*, and *B03* bytes in it. This is followed by *Word 1*, *Word 2*, and so on.

Continuous unaligned stream style

In a continuous unaligned stream, only the data byte and position types can be used, with the condition that position bytes can only be used at the beginning of a data packet, at the end of a data packet, or on both. The following diagram shows a continuous unaligned stream transfer, where the interface is 4 bytes wide:

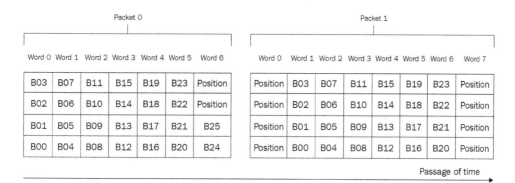

Figure 3.18 – AXI Stream protocol continuous unaligned stream style example

In the preceding diagram, *Word 0* is the first four bytes to be transferred across the bus from the master to the slave and it has *B00*, *B01*, *B02*, and *B03* bytes in it. This is followed by *Word 1*, *Word 2*, and so on. The first packet has a two-byte position type at the end, whereas the second packet has the byte position bytes type both at the beginning and the end of the packet.

Sparse stream style

In a sparse stream, there are data byte and position byte types. The pattern of the data bytes and position bytes should be the same in all packets. The following diagram shows a sparse stream transfer, where the interface is 4 bytes wide:

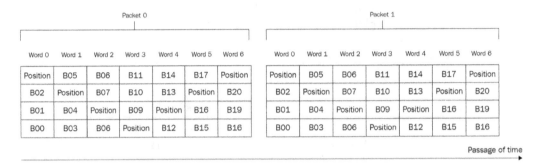

Figure 3.19 – AXI Stream protocol sparse stream style example

AXI Stream bus example system implementation

The AXI Stream bus is used in many designs to solve the issue of having to go through the main system interconnect to exchange data or control information between two specific IPs. For example, the ARM GIC-600 interrupt controller uses the AXI Stream bus to distribute interrupts from the GIC-600 to the CPU cluster, such as the ARM Cortex-73. The following diagram illustrates this connectivity:

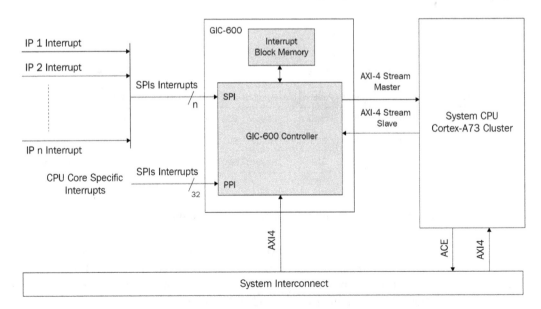

Figure 3.20 – System example of using the AXI stream bus – the ARM GIC-600

ACE bus protocol overview

The **ACE** bus protocol, also known as **AXI with Coherency Extensions**, was first introduced by ARM in AMBA4. It is an AXI-4 protocol with support for cache coherency that's implemented in hardware between different masters, including caches. This section provides an overview of the system cache coherency concepts that will be covered in more detail in the last section of this chapter. We will also explore the ACE protocol, its evolution from AMBA4 to AMBA5 standards, and its added features and mechanisms. We will understand this bus's supported transactions, signaling, and application use cases. We will provide an example system implementation that uses ACE as a bus interface.

ACE bus protocol evolution

The ACE bus protocol is in its second revision as part of the AMBA5 standard. It was first released in AMBA4 as a means to connect masters that include caches and allow the data exchanges to be cache coherent without any software cache maintenance operations. Performing these cache maintenance operations by the hardware interconnecting these masters provides an advantage in terms of processing performance and flexibility in parallel processing software architectures. This is because it makes lock-free programming easier to implement in multi-core and multiprocessor SoCs. In the ACE protocol, a cache line can be in one of the five states supported under ACE. Each state has an associated action when it is accessed under ACE. This extra information and actions require an extra set of signals and channels to extend the AXI-4 protocol to evolve and become an ACE protocol. It also adds support for barrier transactions to enforce transaction completions ordering in the interconnect, as well as support for **Distributed Virtual Memory** (**DVM**) to manage virtual memory addressing at the interconnect level. It extends it to the outside of a specific CPU cluster. AMBA5 standard reorganized the ACE protocol by defining three Lite versions: ACE5-Lite, ACE5-LiteDVM, and ACE5-LiteACP. Barrier transactions support was removed from ACE5 and ACE5-Lite. AMBA5 also removed all references to the low power interface in prior standards as it is now grouped under another protocol, known as the AMBA Low Power Interface specification. In this book, we will focus more on the ACE base protocol, ACE-4, as it is used within the Xilinx Zynq UltraScale+ SoC devices. These are part of the SoCs we will build projects for in the third part of this book.

ACE bus characteristics

As mentioned previously, the ACE protocol is AXI-4 at the base of the specification and provides cache coherency hardware support, which is defined between masters. It also includes and uses the caches of shared memories among them. ACE-4 allows writes to be made to the same memory address by two different masters. These are to be viewed in the same order by all the observing masters that share the same address location of the memory. The ACE-4 protocol implements system-level coherency, where the design can specify the ranges of coherent memories, the memory controllers that support the coherency extensions, and how software can communicate between the elements of the system coherency domain. ACE-4 also provides an abstraction protocol that allows coherency to be implemented between caches that use different coherency protocols. We will introduce coherency protocols in detail in the last section of this chapter. It also allows IPs with different cache line characteristics such as granularity to interoperate coherently. The following diagram provides a simple abstracted example of an ACE-4-based SoC, which should help us study its characteristics:

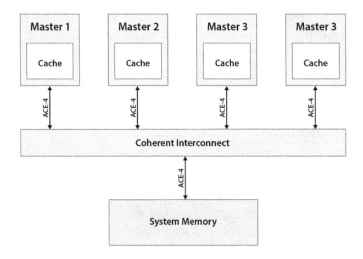

Figure 3.21 – ACE-4 based interconnect simple system example

Cached copies of data from the shared system memory are mapped to all the four master's address spaces as cacheable regions. These can reside in any of the five states in any local cache of one or more master IPs participating in the SoC coherency domain. ACE-4 guarantees that all the masters in the SoC have visibility of the latest value with the data stored in the memory location. This is done by enforcing that only one copy exists (in all the local masters' caches and the system memory) whenever a write operation targets that shared memory location. Once the write operation to the shared memory location is complete, all the participating masters will obtain the latest copy of the data held in the memory location that was targeted by the last write operation. This mechanism allows multiple copies of the data, stored in the shared system memory location as coherent, to exist in multiple locations throughout the caches and the system memory at the same time. As in any efficient coherency protocol, ACE-4 does not require an automatic and immediate write-back of recently updated data in a given cache to the system memory. This is only necessary when a memory location is no longer stored in a *sharable* cache as a result of an explicit software cache maintenance operation or a hardware speculative cache operation on its location, which is elected as a victim under the applied cache hardware manager algorithm. The ACE-4 protocol allows the master components to find out whether a cache line is unique to its cache or if it also exists in another master's cache. In the case of uniqueness, the master can change the content of its cache line without broadcasting the change to all the other participating masters. If it isn't unique, then the master must notify them.

The following table lists all the cache states defined in the ACE-4 protocol:

Cache State	Description
Invalid (I)	The cache line is not present in this specific cache.
UniqueClean (UC)	The cache line is only present in this cache and it still matches the content of the memory location. This allows the owning master to write to this cache line without broadcasting the information to other masters in the system.
UniqueDirty (UD)	The cache line is only present in this cache, and it has been modified vis-a-vis the memory location, which allows the owning master to write to this cache line again without broadcasting the information to other masters in the system.
SharedClean (SC)	The cache line might be present in other caches. It is not known if it still matches the system's memory. However, this master is not responsible for updating it in the system memory, so they must notify any other masters before writing to their cache line while in this state.
SharedDirty (SD)	The cache line might be present in other caches, and it has been modified vis-à-vis the system memory. This master is responsible for updating it in the system memory and must notify any other masters before writing to their cache line while in this state.

Table 3.5 – ACE-4 protocol cache line states description

The ACE-4 protocol states the following rules regarding the cache line states:

- A cache line that is in a unique state can only be resident in one cache in the entire SoC.

- A cache line must be held in a shared state as soon as it is present in more than one cache in the SoC.

- A cache that obtains a new copy of a cache line causes another cache holding the same cache line copy in a unique state to become shared.

- When a cache evicts a cache line, there's no need to inform other cache line holders. This could cause a shared cache line state to only be in one cache.

- A cache line that has been changed vis-à-vis the system memory. It should be in a dirty state.

- A cache line that has been changed vis-à-vis the system memory and is in more than one cache in the SoC should be in a dirty state in only a single cache.

ACE bus interface signals

ACE-4 modifies the read address, write address, and read data channels of AXI-4 by adding the signals listed in the following table. The write data and write response channels are the same in both AXI-4 and ACE-4:

AXI-4 Channel	Added Signal	Description
Read address	ARDOMAIN[1:0]	Driven by the master interface, this indicates the shareability domain of a read transaction as being either Non-Shareable (0b00), Inner Shareable (0b01), Outer Shareable (0b10), or System (0b11).
	ARSNOOP[3:0]	Driven by the master interface, this indicates the read transaction type. There are five groups of transaction types: Non-snooping, Coherent, Cache maintenance, Barrier, and DVM. Each group defines many read transaction types.
	ARBAR[1:0]	Driven by the master interface, this defines the barrier transactions for the read address channel as Normal access, respecting barriers (0b00), Memory barrier (0b01), Normal access, ignoring barriers (0b10), or Synchronization barrier (0b11).
Write address	AWDOMAIN[1:0]	Driven by the master interface, this indicates the shareability domain of a write transaction as being either Non-Shareable, Inner Shareable, Outer Shareable, or System.
	AWSNOOP[2:0]	Driven by the master interface, this indicates the write transaction type. There are four groups of transaction types: Non-snooping, Coherent, Cache maintenance, and Barrier. Each group defines many write transaction types.
	AWBAR[1:0]	Driven by the master interface, this defines the barrier transactions for the write address channel as Normal access, respecting barriers (0b00), Memory barrier (0b01), Normal access, ignoring barriers (0b10), or Synchronization barrier (0b11).
	AWUNIQUE	Driven by the master interface, this is only required by IPs that support WriteEvict-type transactions.

AXI-4 Channel	Added Signal	Description
Read data	RRESP[3:2]	Driven by the slave interface, this extends the AXI-4 read response by two bits to provide the Shareable read transaction completion information. RRESP[2] is defined as the PassDirty bit, while RRESP[3] is defined as IsShared.

Table 3.6 – ACE-4 channels based on the AXI-4 channels description

ACE-4 adds three extra channels to the AXI-4 channels, which are the snoop address channel, the snoop response channel, and the snoop data channel. The snoop address channel provides the value of the snooped address, as well as the associated control information. The snoop response channel is a way for the snooped master to complete the snoop address request and it indicates if a data transfer is going to follow on the snoop data channel. The following table provides an overview of the snoop channel topology:

ACE-4 Channel	Added Signal	Description
Snoop address	ACVALID	Driven by the slave interface, this indicates that the values on the snoop address and the control signals are valid.
	ACREADY	Driven by the master interface, this indicates that the snoop address and control signals can be accepted within the current clock cycle.
	ACADDR[ac-1:0]	Driven by the slave interface, this is the snooped address. Its width should be the same as the read and write address channels.
	ACSNOOP[3:0]	Driven by the slave interface, this indicates the snoop transaction type.
	ACPROT[2:0]	Driven by the slave interface, this indicates the security level of the snoop transaction. Only ACPROT[1] has a meaning in the snoop channel.
Snoop response	CRVALID	Driven by the master interface, this indicates that the snoop response from the master is valid.
	CRREADY	Driven by the slave interface, this indicates that the interconnect is ready to accept the snoop response from the master.
	CRRESP[4:0]	Driven by the master interface, this indicates the snoop transaction's response.

ACE-4 Channel	Added Signal	Description
Snoop data	CDVALID	Driven by the master interface, this indicates the validity of the snoop data back from the master.
	CDREADY	Driven by the slave interface, this indicates that the interconnect slave interface is ready to accept the response data from the master.
	CDDATA[cd-1:0]	This is driven by the master interface and indicates the snoop data that's gone from a master back to the interconnect. "cd" is the width of the snoop data bus.
	CDLAST	This is driven by the master interface and indicates the snoop data transaction's last transfer.

Table 3.7 – ACE-4 specific snoop channels description

ACE bus-supported transactions

ACE-4 protocol defines seven types of transactions:

- **Non-snooping transactions**: These are used to access locations in the address space that are defined as *non-shareable* or *device* – that is, they can't be in another master's cache. To gain access to these locations, the master can use ReadNoSnoop or WriteNoSnoop.

- **Coherent transactions**: These are used to perform coherent access via a vis other master caches – that is, the accessed locations are shareable locations that can potentially reside in the cache of another coherent master in the SoC. To read data from a coherent cache line, the master can issue ReadClean, ReadNotSharedDirty, and ReadShared. To perform a write transaction to shareable locations, a coherent master can use ReadUnique, CleanUnique, and MakeUnique. To write to shareable locations when no cached copy is required, the master can use ReadOnce, WriteUnique, and WriteLineUnique.

- **Memory update transactions**: To update the system memory, the master can use the WriteBack transaction to write back a dirty line of the cache to memory and free it. However, to retain a copy of the cache line, the master should issue a WriteClean transaction. A WriteEvict transaction can be used in a multi-layer cache hierarchy to free a cache line and write it into a lower cache level without updating the system memory. *Evict* is used to broadcast the address in a cache line of a master to other masters without writing any data back to the system memory.

- **Cache maintenance transactions**: These are used by masters in a coherency domain to broadcast cache operations to access and maintain the caches of other coherent masters in the coherency domain. The cache maintenance transactions are CleanShared, CleanInvalid, and MakeInvalid.

- **Snoop transactions**: These use the snoop address, snoop response, and snoop data channels. Snoop transactions are a subset of coherent transactions and cache maintenance transactions.

- **Barrier transactions**: These are used to implement the transactions ordering rule. ACE-4 supports the memory barrier and synchronization barrier types.

- **Distributed virtual memory transactions**: These are used to maintain the system's virtual memory and share information between the elements of the SoC participating in its implementation. It usually includes a **System Memory Management Unit (SMMU)**.

The ACE-4 coherency extension is a complex topic on its own. This section has only provided an overview of it and how its addition makes coherency possible between different participating masters. You are encouraged to consult the ARM ACE-4 specification for more details: `https://developer.arm.com/documentation/ihi0022/e?_ga=2.67820049.1631882347.1556009271-151447318.1544783517`.

ACE bus example system implementation

A good example of using the ACE-4 protocol as the main bus protocol of an SoC is illustrated by the Zynq UltraScale+ SoC interconnect of the APU unit. It includes a **Cache Coherent Interconnect (CCI)** and an SMMU:

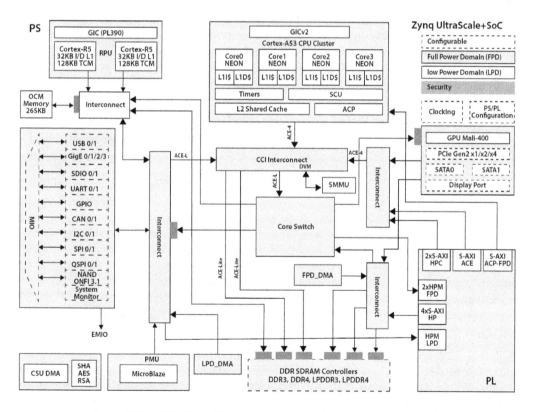

Figure 3.22 – ACE-4 based interconnect system implementation example

OCP interconnect protocol

OCP is an industry standard that provides the specification of a socket-based bus interface that can be used in an interconnect for modern SoCs. It is developed and maintained by the OCP Working Group. The OCP specification is available from Accellera for free and requires the user to accept the Accellera OCP Specification License. Other materials that are provided with the OCP protocol can be found under the Apache 2.0 license. To download the OCP specification, go to `https://www.accellera.org/downloads/standards/ocp/ocp-license-agreement`.

This section only provides a brief introduction to the OCP protocol so that you can gain a high-level understanding of its mechanics and characteristics, as well as compare it to AMBA-based bus protocols such as AXI-4 or ACE-4. However, you are encouraged to study the relevant sections of the specification to further your knowledge of the protocol and its implementation details.

OCP protocol overview

The OCP protocol is a high throughput and high-speed IP interface. It can be implemented in systems that use an SoC interconnect that is not based on the OCP protocol, so long as bridging is performed correctly to and from the OCP interfaces. It helps in overcoming the time-to-market challenges in the SoC's design by providing an industry-standard bus protocol that can cohabit easily with other SoC bus protocols and system interconnects. The OCP protocol is flexible as it can be configured to tailor to the system's design needs. Here, the user can choose the options that are needed and discard the features that aren't specified in their design. This makes it an extensible bus protocol in concept, which also helps in reducing the silicon area cost when using it as an IP bus interfacing protocol. In addition to supporting the **Virtual Component Interface** (**VCI**), the OCP protocol includes the sideband control signals and the test signals. The VCI is specified by the **Virtual Socket Interface Alliance** (**VSIA**).

At the time of writing, the current revision of the OCP protocol is revision 3.0, which added coherence extensions to the base protocol, new sideband signals to control the IP connection state of the interface, and an advanced high-speed profile.

OCP bus characteristics

The OCP protocol is a *point-to-point* connection between two IP cores, with one acting as a master and the other as a slave. Communication is initiated by the master and the slave completes it by accepting the write data from the master or responding with the read data requested by the master in the command. For a data exchange to occur between the two communicating IPs, the master puts the command, the control information, and data in a write scenario on the OCP bus. This goes through an interface block that acts as a slave on the SoC interconnect, which may bridge it to another protocol if the SoC interconnect uses a different one. Then, the end target receives the initial command and the associated information and potential write data. The OCP protocol is flexible and provides a synchronous handshaking signal that allows a pipelined or multi-cycle access model to be used between the master and the slave. It is also worth noting that the OCP protocol is connection-oriented, which is why it's called a socket interface. Here, the IP can connect and disconnect from the bus infrastructure if the SoC interconnect supports this feature.

OCP bus interface signals

OCP interface signals are divided into three categories: *dataflow signals*, *sideband signals*, and *test signals*.

There are three functional sets within the dataflow signals: the *request signals* set, the *response signals* set, and the *data handshake* set. The dataflow signals can also be divided into five groups: *basic* signals, *simple* extensions, *burst* extensions, *tag* extensions, and *thread* extensions.

Only a few basic dataflow group signals are mandatory for an OCP interface configuration. The optional signals can be used to support the need for the communicating OCP interface. The sideband and test signals are all optional. The OCP has a single clock to which all the interface signals are synchronized. The following diagram illustrates the OCP interface signals grouping:

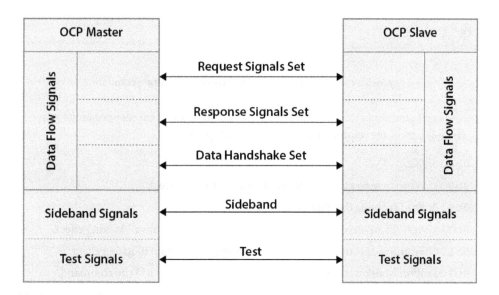

Figure 3.23 – OCP interface signals grouping

For details of all the signals that form the OCP bus, please consult the OCP specification at https://www.accellera.org/downloads/standards/ocp/ocp-license-agreement.

OCP bus-supported transactions

The following diagram shows the OCP protocol's layout and how it is implemented using the predefined groups and sets of signals:

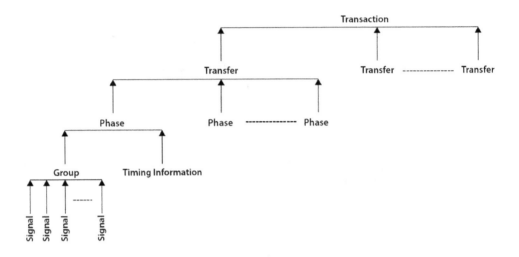

Figure 3.24 – OCP protocol elements hierarchical representation

The transfer commands indicate the types of OCP transactions. These are encoded in the MCmd[2:0] signal, which is part of the request group of signals, as follows:

- 0b000: *Idle*.
- 0b001: *Write* transfers the data from the master to the slave.
- 0b010: *Read* requests the data by the master from the slave.
- 0b011: *ReadEx* is paired with a *Write* or *WriteNonPost* and has a blocking effect.
- 0b100: *ReadLinked* is used with *WriteConditional* and has no blocking effect.
- 0b101: *WriteNonPost* is used to request the slave not to post a Write command.
- 0b110: *WriteConditional* is used to write only when there is a reservation on the write address set. If not, it fails.
- 0b111: *Broadcast* is used when the addressed location may map to more than one slave.

ReadExclusive, ReadLinked, and WriteConditional have similar effects on some processor semantics that are used to synchronize multiple CPU cores in an SoC.

The OCP protocol also supports burst operations to achieve high system throughput with minimum overhead. OCP burst operations may include addressing details for each transfer command or just one for the full burst transfer.

The OCP protocol uses tags to relax operation ordering that belongs to the same master, so operations with different tags may be treated with a relaxed ordering in the completion phase by the slave.

DMA engines and data movements

Modern SoCs and high-performance IPs include DMA engines to offload data movement from the CPUs and perform it as desired by the software, with the option to notify the CPU once the operation is completed. There are two categories of DMA engines: DMA engines that are included within an IP, and central DMAs that are standalone and connected to the SoC interconnect like any other SoC IP.

IP-integrated DMA engines overview

IP-integrated DMA engines act as data movers on behalf of the IP and the CPU, so the CPU won't be needed to copy data from system memory to the IP's local storage or vice versa. Rather, when programmed and armed, once the data is needed or received, the DMA engine autonomously performs the data transfer from source to destination. Then, its control hardware notifies the CPU when the operation has finished executing, usually via an interrupt. The following diagram illustrates an IP-integrated DMA engine:

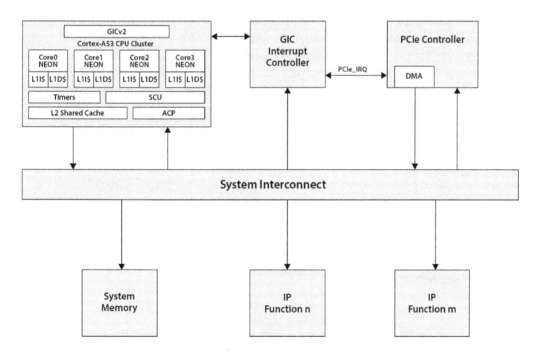

Figure 3.25 – IP-integrated DMA engine example

IP-integrated DMA engines topology and operations

Most high-speed IPs include a DMA engine that has multi-channels. They are also full-duplex so that data movement isn't the SoC bottleneck for meeting system performance requirements. The topology of the DMA engine to include many channels that are capable of reading and writing simultaneously is to be able to service the data movement needs of the IP itself. High-performance communication IPs, such as the PCIe controllers and the multi-gigabit Ethernet controllers, can move a high quantity of data in both the transmit and receive directions. They usually service many software threads running on different CPU cores. These use case scenarios require an efficient and low overhead data movement within the SoC between the system memory and these controllers. The integrated DMA engines within these controllers are the appropriate solution to meet these needs.

Integrated DMA engines, in contrast to central DMA engines, have the advantage of moving the data directly from system memory to the controller or vice versa. Consequently, the data reaches its destination in half the required time it may need when using a central DMA engine, where data needs to be copied to the central DMA engine buffer from the source, then written out to the destination.

IP-integrated DMA engines are capable of moving data from/to regular memory with sequential addressing and from/to peripherals (where a FIFO is usually used as storage) and the address of the source or destination is the same. To support peripherals with integrated FIFOs (and if not managed by the IP peripheral itself behind a bus interface), some integrated DMA engines use handshaking signals to manage the communication directly with the FIFO. These DMA engines aren't generic and are usually designed to support a specific hardware interface. Moving data from regular memory is generic in the sense that the DMA engine will interface to the system interconnect via, for example, one of the covered AMBA or OCP bus protocols and then perform the operations using these bus protocols.

The data movements to be performed are seen as operations that are executed by the DMA engine. These operations are performed by the DMA hardware state machine using operands that are specified via registers local to the DMA engine. This set of operands is called the DMA operation *descriptor*. Among the required operands in a DMA descriptor for a read/write from/to system memory where the data is stored/to be written in a contiguous physical buffer, we can find the following:

- The source/destination address of the data
- The size of the data transfer
- Control information
- Start/stop

The DMA engine then stores the results of the operation, such as a *progress watermark*, error status, and completion status, in a set of registers that software can read when appropriate.

There are also DMA engines capable of transferring complex structures of the data from the system memory to its local storage without any data rearrangements. These complex data structures are usually data buffers scattered around physical buffers of different sizes in the system memory. Some DMA engines can perform such transfers using a *linked list* of the aforementioned DMA descriptors, where the DMA engine will read them in the execution stage sequentially. Once the executed list reaches the last entry, it will point to the next list. This list is usually a contiguous number of descriptors; however, the last element of the list is just a pointer to the next list of descriptors. This result in the software needing to define the linked list of descriptors in the system memory, and then provide the address of the first element of the linked list of descriptors to the DMA engine to load before launching the DMA operation. The process of loading the linked list of DMA descriptors is often called *pre-fetching* as it is an operation that's performed while a DMA operation of a previous sequence is ongoing. The operands for such DMA operations are as follows:

- The type of the descriptor is a data buffer descriptor or a pointer to the first element of the next linked list.
- The source address of the data is a data buffer or the location of the next first element of the next list.
- The size of the data to transfer if it's not a pointer.
- Control information.
- Start/stop.

Every IP-integrated DMA engine may have a specific way of implementing this to interface with the software, but the general concept is pretty much the same. The following diagram illustrates the DMA descriptors linked list concept:

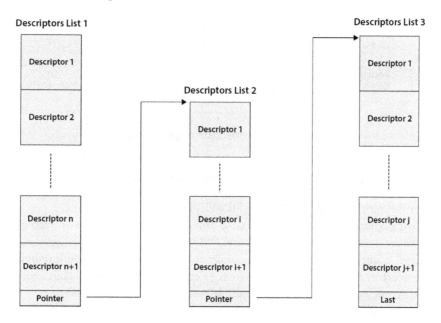

Figure 3.26 – IP integrated DMA descriptors linked list example

Standalone DMA engines overview

Standalone DMA engines are often called **central DMAs** as they are central in the SoC interconnect and can move data from anywhere in the SoC-mapped address space to anywhere else, as far as they are allowed to do so by the SoC interconnect and any security policy that may restrict access to some regions of the address space.

Like the IP integrated DMA engine, a central DMA IP autonomously performs the data transfer from source to destination. It control hardware proceeds to notify the CPU via an interrupt once the required transfer operation has finished. The following diagram illustrates the notion of central DMA engines:

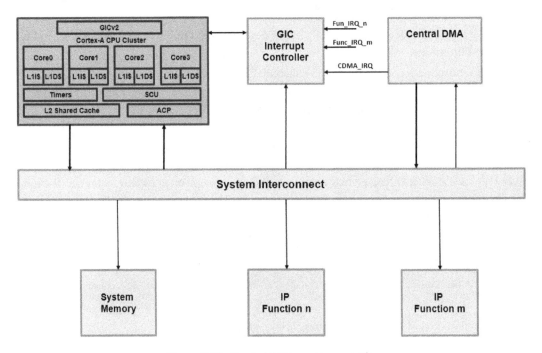

Figure 3.27 – Central DMA engine example

Central DMA engines topology and operations

Most modern SoCs include a central DMA engine with multi-channels that are full-duplex to accelerate the data moving operations on behalf of the CPU and IP peripherals that have no integrated DMA engine. There are many topologies of the central DMA engines that vary in design complexity and features, as well as silicon area cost. However, usually, central DMA engines include multiple channels capable of concurrent read and write transfers from different sources and destinations either by sharing their master interface or by implementing multiple masters.

Central DMA engines, as already mentioned and in contrast to IP integrated DMA engines, need to copy the data from the source to the central DMA buffer, then copy the data from the central DMA buffer to the destination. Consequently, the data reaches its destination in twice the required time it may need when using an IP-integrated DMA engine, where data is transferred directly from the IP to the destination or vice versa.

A central DMA engine can move data between different types of sources and destinations, as follows:

- Memory to memory
- Memory to FIFO
- FIFO to memory
- FIFO to FIFO
- **Scatter Gather (SG)**

A Scatter Gather DMA operation is a specific type of data transfer from memory to memory where the source of the data and its destination is formed by non-contiguous buffers in the physical memory address space.

Similar to an integrated DMA engine, the data movements to perform are seen as operations that are executed by the central DMA engine. These operations are performed by the DMA hardware state machine using operands that are specified via registers local to the DMA engine. This set of operands is called the DMA operation *descriptor*. Among the required operands in a DMA descriptor for a read/write from/to system memory where the data is stored/to be written in a contiguous physical buffer, we can find the following:

- The source address of the data
- The destination addresses of the data
- The size of the data transfer
- Control information
- Start/stop

Then, the central DMA engine stores the results of the operation, such as a progress watermark, error status, or completion status, in a set of registers that software can read when appropriate.

Like the IP-integrated DMA, central DMA engines are also capable of moving complex data structures from source to destination using the linked list of DMA descriptors concept explained in the previous section.

Data sharing and coherency challenges

Modern SoCs are constructed using multiple data compute engines such as CPUs, GPUs, custom hardware accelerators, and high-performance IP peripherals with integrated DMA engines. They process data that is shared laterally and passed from layer to layer as these different processing engines cooperate. These complex topologies make the system design more challenging in terms of making sure that the data is safe as it is accessed, used, and updated while also ensuring it is coherent when these processing engines make efforts to optimize the data access, such as by using integrated caches. Accessing data concurrently and safely means accessing it atomically and coherently without paying a high penalty in terms of software management and using prohibitive locking mechanisms. We want the system architecture to be lock-free and rely on the hardware to provide the optimal mechanisms to implement these data protection mechanisms.

Data access atomicity

To make sure that the data is accessed atomically in an SoC with multi compute nodes sharing it, the SoC system architecture should provide exclusive access primitives that cooperating compute elements can use to protect access to shared data. The atom of data access is specific to the application and can be a simple data type, such as an integer, or a custom-defined data structure that the application software wants to be accessible fully or partially to a single requester at a time. Traditionally, hardware-based primitives were used, such as hardware mutexes, and these can still be used in today's SoCs. ARM also defines a whole end-to-end paradigm – that is, from the software layer to the local exclusive monitors within the cluster and the global exclusive monitors within the system memory controllers. These monitors are hardware-based state machines are implemented at the bus interface level of the CPU cluster for the local exclusive monitors or the bus side interface of the memory controller for the global exclusive monitors. They observe access sequences to memory locations where the semaphore-based primitives are located. This means that software threads can use these semaphores as a protective mechanism to access shared data between software threads running on the same CPU core, between software processes running on different CPU cores within the same CPU cluster, and between hardware or software threads running outside of a specific CPU cluster. Using hardware-based access protection primitives or exclusive (local and/or global) monitors is a choice that can be made based on the system's sharing and performance requirements. System profiling should help with making the optimal choice, though using a mixture of both is usually the solution.

> **Information**
>
> If you wish to learn more about the global exclusive monitors topic, you are invited to study the ARM material at https://developer.arm.com/documentation/dht0008/a/arm-synchronization-primitives/exclusive-accesses/exclusive-monitors?lang=en.

Cache coherency overview

Cache coherency isn't a new topic in the SoC design since it is the basis of building a multi-core cache-coherent CPU cluster. However, it now extends beyond the CPU cluster and reaches the SoC level via bus protocols such as ACE and OCP.

Cache coherency protocols overview

The following diagram illustrates the concept of sharing data between two CPUs and integrating caches in them. The shared data is cached from system memory. There is a need for a cache coherency protocol within the CPU cluster, a bus interface with coherency support such as ACE-4, and an interconnect with coherency support to allow these CPUs to share data coherently without costly (in terms of performance) software cache management being required:

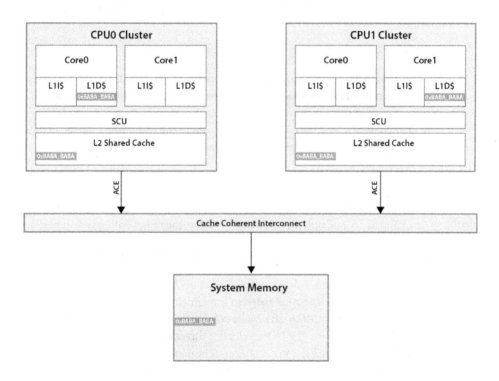

Figure 3.28 – Cache coherency support infrastructure

If CPU1 modifies the content of the cache line that includes the 0xBABA_BABA location from system memory, then the **Snoop Control Unit (SCU)** will make sure that this access is done coherently since it has all the system coherency infrastructure to do so. Any CPU core from a CPU cluster can access shared data that is also cached coherently without the need to perform any software cache maintenance operations. This is because the hardware will take care of that autonomously.

Implementing cache coherency in CPUs

In a CPU cluster that's integrating multiple cores, each with level 1 caches and at least another higher-level cache such as a common L2 cache, there is usually an SCU that orchestrates data accesses and makes sure that data is shared efficiently between the different CPU cores within the cluster, as well as shared coherently. The SCU implements a coherency protocol, which is usually a derivative of the *MESI* cache coherence protocol. Let's look at what the MESI cache coherency protocol stands for:

- **M**: This stands for *Modified* and means that the cache line has been modified vis-à-vis the system memory.

- **E**: This stands for *Exclusive* and means that the cache line is exclusively present in the current cache.

- **S**: This stands for *Shared* and means that the cache line is present in at least another cache with this one.

- **I**: This stands for *Invalid* and means that the cache line is invalid.

Modified, Exclusive, Shared, and Invalid are four states under which, and at any time, a given cache line of any CPU core can be held. The modern multi-core CPU clusters have at least another state added to the MESI protocol to support useful cache coherency protocols, such as *Owned*. The protocol is called *MOESI*. Owned means that the cache controller where the cache line is in the Owned state and is responsible for updating the next-level cache or main memory.

With these five states, most of the cache controllers can, alongside the SCU, implement efficient cache coherency protocols and share data between the caches of the CPU cores within the cluster without having to go via memory. This reduces the data sharing latency, which, in turn, reduces power consumption by avoiding heavy bus and external memory traffic, thus augmenting system performance.

Extending cache coherency at the SoC level using the accelerators port

Some CPUs such as the ARM Cortex-R and Cortex-A have an **Accelerator Coherency Port** (**ACP**) that extends the cache's coherency. The ACP helps in implementing an I/O-coherent topology that allows hardware accelerators and the CPU to share data coherently without software cache management assistance. However, this is only true in the I/O accelerator direction. This means that if the hardware accelerator integrates a cache, the CPU can't access system memory that's been cached within the hardware accelerator coherently. Therefore, it is important to understand how the ACP port operates in both the Cortex-A9 CPU cluster within the Zynq-7000 SoC and the ACP port of the Cortex-A53 CPU cluster of the Zynq UltraScale+ SoC.

When using the ACP port to extend I/O coherency with an accelerator, for example, care must be taken to ensure that the data access patterns and transaction types that are used by the accelerator master are supported by the ACP port. Not all the transactions that are supported by the bus protocol and used by the ACP port are permitted and there are limitations that the user needs to be aware of. These can be found in the CPU **Technical Reference Manual** (**TRM**).

When access is gained through the ACP, usually, if there is a hit in a CPU core cache, then the transaction is performed from the cache (be it read or write). However, if there is a cache miss, then the transaction is fulfilled from system memory, so there is a price to pay here in terms of added cycles of latency and performance sharing the CPU cluster's throughput to the system memory in case of a transaction miss. The ACP port in certain CPUs allows allocation on miss, which can be used when there is a follow-up need by the CPU core in that cluster regarding that data in the future. However, at the same time, cache allocation from the ACP may evict warm data that the CPU may potentially need in the future and probably more critically. Consequently, when using the ACP and setting its behavior on misses, profiling with this information should help you make the appropriate decisions.

Extending cache coherency at the SoC level using the ACE-4 and CCI

As shown in *Figure 3.28*, coherency can be extended beyond the CPU cluster by using a bus protocol such as ACE-4 and an interconnect that supports the ACE-4 coherency protocol, such as the ARM CCI-400. The SCU within the CPU cluster will perform the necessary cache lines snooping and cache line management information broadcasting to make sure that a central cached memory region, by more than one CPU cluster with integrated caches, is shared coherently between all the CPU cores of this SoC example.

The following diagram focuses on the Zynq UltraScale+ SoC coherency features, including the Cortex-A53 cluster-integrated coherent caches, along with the SCU, its ACP port, and its ACE-4 interface. These can be used to access some of the system ports via the **cache-coherent interconnect** (**CCI**):

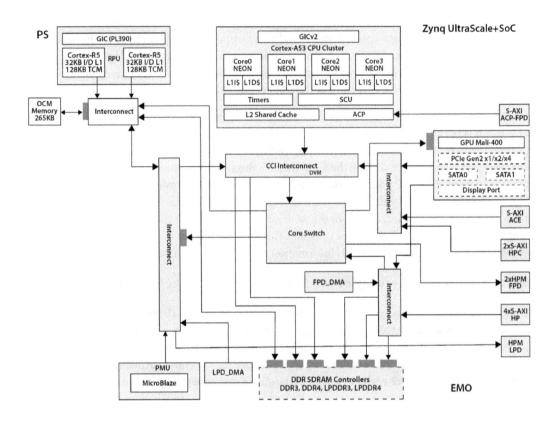

Figure 3.29 – SoC level coherency implementation example

Some FPGA masters can also access the PS block coherently via the S-AXI ACE-4 port.

Summary

In this chapter, we introduced buses and interconnects and their crucial roles in putting modern SoCs together. We explored the defining functional features of buses and interconnects and the background behind them. We also looked at the ARM AMBA standards by revisiting their historical evolution and how SoC design complexity also required an evolving standard that can accommodate and help with designing higher performance and features-rich SoCs. We explored all the relevant bus protocols that make up the AMBA standard, their features and characteristics, and what makes them suitable for a specific connectivity need. We also looked at example implementations to get a feel of how these buses are used in modern SoCs. Then, we covered the OCP standard, its bus characteristics and features, and what makes it appealing for many application domains in the SoC design space. We compared the AMBA standard and the OCP bus protocol and how they can be bridged to accommodate mixed standard complex interconnects. Continuing to look at the interconnect and data exchange SoC functionality, we looked at DMA engines, the supported data exchanges and transaction types, and how a data movement from source to destination can be defined, launched, and executed by a DMA engine on behalf of a requesting master on the SoC. We also covered DMA engine software to hardware interfacing. We concluded this chapter with an overview of the data-sharing challenges in modern complex SoCs and their data coherency and atomic access, including how these challenges can be dealt with efficiently to avoid sharing data in high-performance SoCs becoming a system performance bottleneck.

The next chapter will continue to address the SoC design and its architecture fundamentals by looking at the SoC's off-chip connectivity to other high-speed devices using mediums such as PCIe and Ethernet.

Questions

Answer the following questions to test your knowledge of this chapter:

1. Describe the differences between a simple bus and a Network-On-Chip. What is a multi-threaded master?

2. Which bus protocols are included in AMBA4? Describe the main characteristics of each.

3. Which states are part of the ACE-4 coherency protocol? Describe each.

4. What are the differences between the OCP bus and the AXI-4 bus? What are the similarities between them?

5. How many types of DMA engines are there? Describe the main differences between them.

6. What is a DMA descriptor? What are its fields?

7. What is a Scatter Gather DMA operation?

8. What is a linked list of descriptors? What is the key element in it?

9. How is cache coherency implemented at the SoC system level?

10. What is atomic data access? How can we implement this between two different CPU clusters?

Connecting High-Speed Devices Using Buses and Interconnects

This chapter begins by giving an overview of the buses and interconnects used off-chip to connect an FPGA-based SoC to other high-speed devices on the electronics board. It introduces the PCIe interconnect, the Ethernet interconnect, and the emerging Gen-Z protocol. It also provides a good overview of the emerging CCIX interconnect protocol and the concept of extending data coherency off-chip by adding protocol layers in the SoC hardware to implement it.

In this chapter, we're going to cover the following main topics:

- An overview of the legacy off-chip interconnects

- An introduction to the PCIe bus

- An overview of the Ethernet interconnects

- An introduction to the Gen-Z bus protocol

- An overview of the CCIX protocol and off-chip data coherency

Legacy off-chip interconnects overview

Connecting devices together using off-chip buses and interfaces is necessary to construct an electronics system made up of many **integrated circuits** (**ICs**). The need to use these buses is to satisfy the requirements of these ICs to exchange data as they collaborate to implement a more complex function in comparison to what they can achieve on their own. The electronics industry has seen the emergence of many standards and protocols to address this need, and the choice of using a specific protocol depends on many factors, such as the physical IO standards, the speed that these buses can run at, the implementation cost, the versatility of the protocol in the industry, the throughput that they can provide, and so on. We distinguish between two main categories of buses in legacy and modern systems: the protocols suited to carry control data and signaling between devices, and the buses that carry high-throughput payload data at a relatively high speed. The legacy buses have several standards and protocols in use, such as the **Serial Peripheral Interface (SPI)**, the **Inter-Integrated Circuit (IIC, but also known as I2C)**, the **Universal Asynchronous Receiver-Transmitter (UART)** interface, and many custom buses implemented using a serial or parallel custom protocol carried over low-speed **General Purpose Input/Output (GPIO)** interfaces. In this section, we will briefly introduce the SPI and I2C buses as some of the most common legacy bus protocols; for a more in-depth study of these bus standards, you are encouraged to examine the respective bus standard specification referenced in their respective section that follows.

SPI overview

The SPI is a simple serial synchronous bus protocol developed by Motorola in the 1980s for connecting a master and a slave device. It is implemented using four signals:

- **SCLK: Serial clock**, a signal generated by the master

- **MOSI: Master out slave in**, an output data signal from the master

- **MISO: Master in slave out**, an output data signal from the slave

- **SS: Slave select**, an output control signal from the master to a specific slave device, acting as a **chip select (CS)** signal

The following diagram illustrates the connectivity between a master and a slave SPI device:

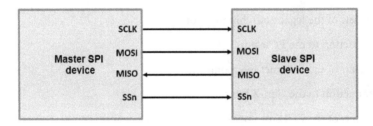

Figure 4.1 – SPI bus connecting a master and a slave SPI device

> **Information**
> Further information regarding the SPI bus protocol is available in the SPI specification, which can be accessed at https://www.nxp.com/files-static/microcontrollers/doc/ref_manual/S12SPIV4.pdf.

When included in an SoC as an external bus interface, it is integrated into the design within a controller that also provides an internal interface usually via the ARM AMBA APB bus connecting it to the SoC. The SPI controller provides a register file via which the SoC CPU can configure the SPI bus operation modes and speeds. The SPI controller usually supports the full-duplex operation by performing simultaneous receives and transmits. It can buffer the read and write operations within a transmit and a receive FIFO. The master clock frequencies are programmable to adapt to the SoC and external device operational needs. The SoC CPU can control the SPI controller through polling or can use interrupts.

Zynq-7000 SoC SPI controller overview

The following diagram provides a micro-architectural view of the APB SPI controller included in the Xilinx Zynq-7000 SoC **processor subsystem (PS)** block:

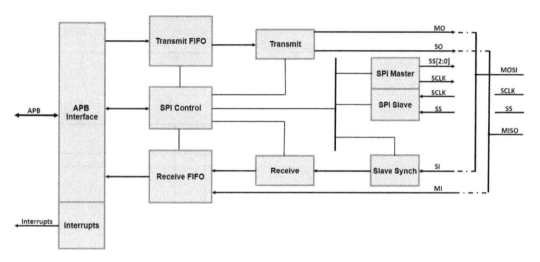

Figure 4.2 – Example SPI controller – the Zynq-7000 SoC

The Xilinx SPI controller can operate either in master or slave mode. In the master mode, the SPI controller can write data out to the slave over the SPI bus or trigger a read operation from the slave. It can interface with up to three slave SPI devices and can distinguish among them using one of the three SS signals (SS[2:0]). Both the SCLK and MOSI are controlled by the master. To write the data from the master to the slave, the SoC CPU writes it first into the transmit FIFO over the APB interface, the SPI controller hardware, then serializes the data out on the MOSI pin. The data write operation is continuous while there is still data in the transmit FIFO of the SPI controller. The read data is received serially via the MISO pin and written by the SPI controller hardware 8 bits at a time into the receive FIFO of the SPI controller. The SCLK frequency can reach 50 MHz in the Zynq-7000 SoC FPGAs, giving a decent data bandwidth for a low-throughput interface.

> **Information**
>
> For further operational and configuration details of the Zynq-7000 SoC SPI controller, you are encouraged to examine section 17 of the *Zynq-7000 SoC Technical Reference Manual*, which can be found at https://docs.xilinx.com/v/u/en-US/ug585-Zynq-7000-TRM.

I2C overview

I2C was invented by Philips in the 1980s; it is also a serial bus protocol using packet switching as a data transport mechanism between the electronics systems' components. It only requires two bidirectional signals:

- **SDA**: Serial data line
- **SCL**: Serial clock

The following diagram illustrates the connectivity between a controller and a target device:

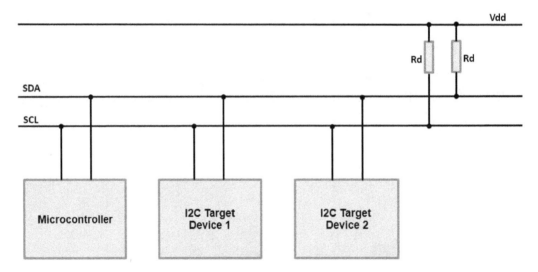

Figure 4.3 – I2C bus connecting a controller and target devices

Every device on the I2C bus is assigned a unique address by the system software. Controllers can operate as transmitters or receivers. The I2C bus protocol allows system topologies to include many controllers on the bus, as it has collision detection and arbitration capabilities that help in preventing data transmission errors if multiple controllers start transmitting at the same time. Data transfer is 8-bit oriented and the throughput in the bidirectional case can achieve up to the following:

- 100 KB in *Standard mode*
- 400 KB in *Fast mode*
- 1 MB in *Fast-mode Plus*
- 3.4 MB in *High-speed mode*

In the 8-bit oriented unidirectional case, the data transfer throughput can reach up to 5 MB in *Ultra-Fast mode*.

The maximum bus capacitance (*400 pF*) and the 7-bit (or in extended mode 10-bit) address space are the only limitations to the number of ICs that can be interconnected using the same I2C bus.

Information

For an in-depth study of the I2C bus standard, you are encouraged to examine its specification available from NXP at https://www.nxp.com/webapp/Download?colCode=UM10204.

When used in an SoC as an external bus interface, the I2C controller usually uses an SoC bus interface such as the APB to connect to the SoC interconnect. The I2C controller provides a register file via which the SoC CPU can configure the I2C bus operation modes and speeds that the I2C controller supports. The master clock frequencies are programmable to adapt to the SoC and external device operation's needs. The SoC CPU can control the I2C controller through polling or can use interrupts.

Zynq-7000 SoC I2C controller overview

The following diagram provides a micro-architectural view of the APB I2C controller included in the Zynq-7000 SoC PS block:

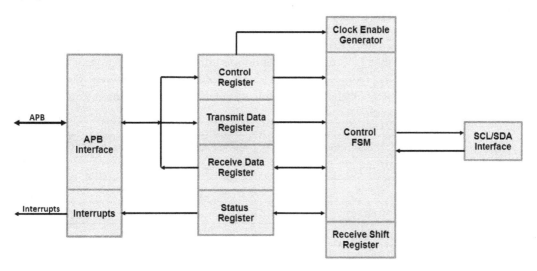

Figure 4.4 – Example I2C controller – the Zynq-7000 SoC

The Xilinx I2C controller can operate either in master or slave mode. In the master mode, the I2C controller can write data out to the slave over the I2C bus or request a read operation from the slave. The Xilinx Zynq-7000 SoC I2C controller supports only the *Standard mode* (100 KB) and the *Fast mode* (400 KB) transfer modes.

> **Information**
>
> For in-depth operational and configuration details of the Zynq-7000 SoC I2C controller, you are encouraged to examine section 20 of the *Zynq-7000 SoC Technical Reference Manual*, which can be found at `https://docs.xilinx.com/v/u/en-US/ug585-Zynq-7000-TRM`.

Introduction to the PCIe interconnect

The **Peripheral Component Interface Express** (**PCIe**) is a high-speed, multi-layer, and serial interconnect protocol. Its predecessor, the **Peripheral Component Interface eXtended** (**PCI-X**), was a parallel interface, but most of its base architectural properties are included in PCIe. The PCIe protocol defines three protocol layers: the transaction layer, the data link layer, and the physical layer. The physical layer uses multi-gigabit transceivers and can communicate at tens of gigabits per second. The physical layer topology is formed of multiple parallel transceivers known as lanes to transport data at a high bandwidth to match the application data transfer rates requirement. There are now many generations of PCIe protocol and all are backward compatible, from **generation 1** (**Gen1**) to **generation 6** (**Gen6**).

Historical overview of the PCIe interconnect

The first PCIe generation, Gen1, was introduced in 2003 as a replacement for the PCI-X parallel interconnect; since then, each subsequent generation was introduced after a few years to basically double the throughput of its predecessor and add features deemed necessary by system applications where PCIe was or can potentially be used. The following table summarizes the PCIe generations' evolution over the last 20 years:

Introduction Year	PCIe Generation	Lane Rate (GT/s)	Throughput (GB/s)
2003	Gen1	2.5	0.250
2007	Gen2	5.0	0.500
2010	Gen3	8.0	0.985
2017	Gen4	16.0	1.969
2019	Gen5	32.0	3.938
2022	Gen6	64.0	7.563

Table 4.1 – PCIe generations' historical evolution

The PCIe Gen1 and Gen2 physical layers use an 8b/10b line encoding, where 8 bits of data are mapped to a 10-bit symbol designed specifically to provide the optimal transmission conditions. These mainly are the balancing of the **direct current** (**DC**) on the differential transmission medium and providing the minimum necessary transitions to recover the clock at the receiver side. Basically, in the multi-gigabit transceivers, the clock is embedded with the data and recovered at the receiver end from the symbol's transitions.

> **Information**
>
> For more details on the PCIe 8b/10b encoding, you can check the PCIe Gen1 or Gen2 specification, which can be found on the PCI-SIG at `https://pcisig.com/specifications`.
>
> Note: you will need to register to gain access to the PCIe base specifications.

In a similar approach, while also seeking to reduce the physical layer encoding overhead, PCIe Gen3, Gen4, and Gen5 physical layers use a 128b/132b line encoding, where 128 bits of data are mapped to a 132-bit symbol. For more details on the PCIe 128b/132b encoding, you can check the PCIe Gen3 specification, which can be found at the PCI-SIG URL provided previously. PCIe Gen6 uses **pulse-amplitude modulation** (**PAM**) to encode the data into the signal combined with a logical data integrity mechanism. The PCIe Gen6 PHY inserts an 8-byte **cyclic redundancy check** (**CRC**) and a 6-byte **forward error correction** (**FEC**) for every 242 bytes of data to make a physical layer FLIT of 256 bytes.

PCIe interconnect system topologies

PCIe interconnect is a point-to-point full-duplex connection, meaning that both ends can transmit and receive simultaneously. The following diagram illustrates this connectivity concept:

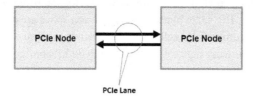

Figure 4.5 – PCIe point-to-point connectivity

The protocol transaction layer, as will be covered in detail in this chapter, allows routing between nodes, which makes PCIe also switchable. The following diagram illustrates the possible system topologies using PCIe as an interconnect that uses a PCIe switch:

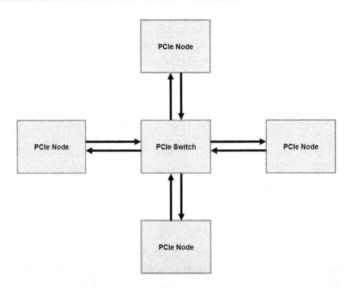

Figure 4.6 – PCIe system using a switch and many PCIe nodes

In a PCIe system, there is always a single upstream device called the PCIe **root complex** (**RC**). The PCIe RC is responsible for setting up the PCIe system and configuring the PCIe devices, also called PCIe **end points** (**EPs**). Each PCIe device has a configuration space that defines many of its operational aspects at runtime. At power up, the PCIe RC runs what is called a PCIe hierarchy discovery and in this process, it reads the PCIe EP configuration space, which allows it to set them up according to the PCIe system needs. It also builds the PCIe tree by allocating to each PCIe device a bus number, a device number, and a function number, forming the device ID of the PCIe EP. Each PCIe EP can have multiple logical functions, up to eight functions, and each logical function has an associated 4,096 bytes of configuration space. The 4,096 bytes of configuration space are divided into two regions: the first 256 bytes is a PCI 3.0-compatible region, and the remaining space is called the PCIe extended configuration space. The following figure illustrates the PCIe EP configuration space:

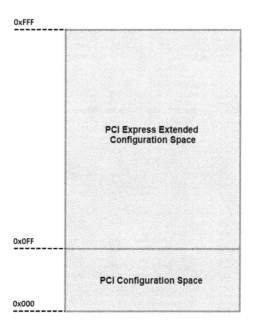

Figure 4.7 – PCIe configuration space layout

The PCIe configuration space, which includes both the PCI configuration space and the PCIe extended configuration space, uses configuration registers that the PCIe RC can read and write to discover the PCIe EP device capabilities and set certain system behaviors. It is the software running on a CPU usually that uses the PCIe RC as a proxy to configure the PCIe EP devices belonging to a certain PCIe hierarchy. Part of the PCIe configuration space is the PCIe configuration space header, of which there are two types: Type 1 for the PCIe RCs, PCIe bridges, and PCIe switches, and Type 0 for the PCIe EPs. The following diagram illustrates the Type 0 PCIe configuration space header:

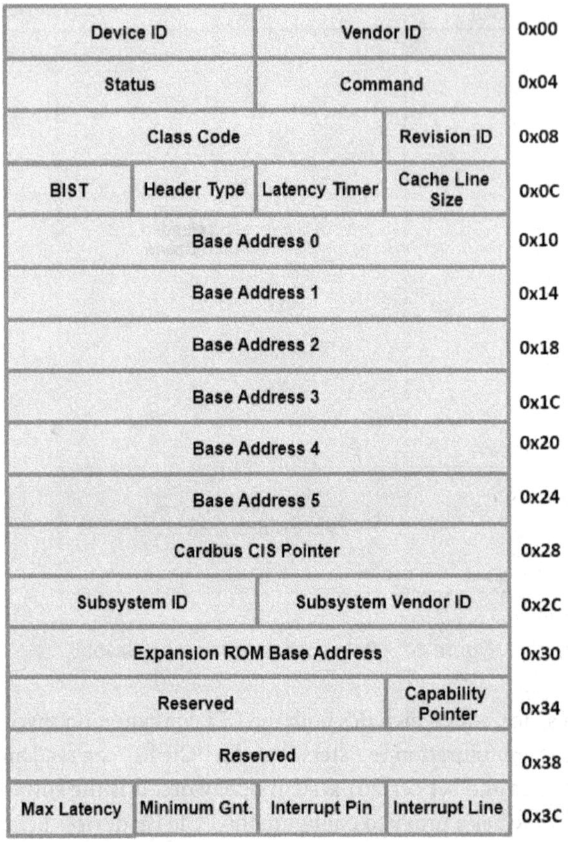

Device ID		Vendor ID		0x00
Status		Command		0x04
Class Code			Revision ID	0x08
BIST	Header Type	Latency Timer	Cache Line Size	0x0C
Base Address 0				0x10
Base Address 1				0x14
Base Address 2				0x18
Base Address 3				0x1C
Base Address 4				0x20
Base Address 5				0x24
Cardbus CIS Pointer				0x28
Subsystem ID		Subsystem Vendor ID		0x2C
Expansion ROM Base Address				0x30
Reserved			Capability Pointer	0x34
Reserved				0x38
Max Latency	Minimum Gnt.	Interrupt Pin	Interrupt Line	0x3C

Figure 4.8 – PCIe configuration space header

The following table defines the Type 0 PCIe configuration space header fields:

Field	Size (B)	Description
Device ID	2	Chipset manufacturer-assigned ID to this PCI/PCIe device.
Vendor ID	2	Chipset manufacturer ID of this PCI/PCIe device.
Status	2	Register for which features are supported by this PCIe EP device and report certain types of errors.
Command	2	This is a bitmask register for enabling and disabling individual features of this PCIe EP device.
Class Code	3	Indicates the type of function this device implements.
Revision ID	1	This field identifies a specific revision of this device.
BIST	1	Used to perform the PCIe EP built-in self-test.

Field	Size (B)	Description
Header Type	1	Indicates the format of the rest of this configuration space header from 0x10 upward, and indicates whether this PCI/PCIe device has multiple functions: • 0x0: This is a general device. • 0x1: PCI-to-PCI bridge. • 0x2: CardBus bridge. When bit 7 of this register is set, then this is a multi-function PCI/PCIe device, otherwise, this is a single-function device.
Latency Timer	1	Indicates this PCIe EP device latency timer in units of the PCI bus clock (33 MHz), that is, in 30 ns units.
Cache Line Size	1	This register doesn't apply to the PCIe devices, it is only for PCI devices and usually matches the PCIe RC CPU cache line size.
Base Address x	4	x = 0 to 5, used to specify the memory and I/O space mapping of the PCIe EP to the PCIe address space.
Cardbus CIS Pointer	4	Used in devices that share CardBus and PCI, it provides a pointer to the card information structure.
Subsystem ID	2	Identifies the card manufacturer.
Subsystem Vendor ID	2	Identifies the card and is assigned by the card manufacturer from the same number space as the device ID.
Expansion ROM Base Address	4	The base address of the expansion ROM.
Capability Pointer	1	Provides a pointer to the first capability of this PCIe EP device, applicable only when bit four of the status register is set to 0b1.
Maximum Latency	1	Provides the access frequency of this PCI device to the PCI bus, expressed in a period of 250 ns units.
Minimum Gnt.	1	Minimum grant: it indicates the length of this device burst period in 250 ns units.
Interrupt Pin	1	Applies only to the PCI devices and is used for the signal-based PCI interrupts (A, B, C, and D).
Interrupt Line	1	Applies only to the PCI devices and is used for the signal-based PCI interrupts (A, B, C, and D).

Table 4.2 – PCIe configuration space Type 0 header fields definition

A simple PCIe-based subsystem is depicted in the following diagram:

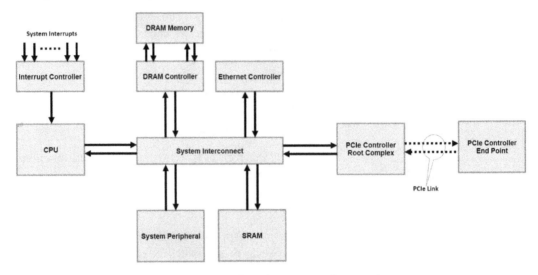

Figure 4.9 – PCIe-based system simple example

The PCIe link can be formed using a single bidirectional PCIe lane, denoted x1; it can also be formed using two lanes (x2), four lanes (x4), eight lanes (x8), sixteen lanes (x16), or even thirty-two lanes (x32). The following table summarizes the PCIe link throughput per link width and speed in GB/s:

PCIe Generation	x1	x2	x4	x8	x16	x32
Gen1	0.250	0.500	1.000	2.000	4.000	8.000
Gen2	0.500	1.000	2.000	4.000	8.000	16.000
Gen3	0.985	1.969	3.938	7.877	15.754	31.508
Gen4	1.969	3.938	7.877	15.754	31.508	63.015
Gen5	3.938	7.877	15.754	31.508	63.015	126.030
Gen6	7.563	15.125	30.250	60.500	121.000	242.000

Table 4.3 – PCIe link bandwidth

PCIe protocol layers

The PCIe protocol defines three stacked layers: the transaction layer, the data link layer, and the physical layer. Each layer has a mixture of metadata and the associated payload data. Each protocol layer packet is encapsulated and transported within the next layer frame that adds to it its own protocol metadata. The following diagram shows the PCIe protocol stacked layers concept:

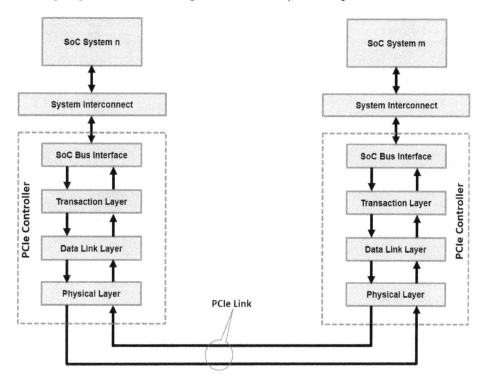

Figure 4.10 – PCIe protocol layers

The PCIe transaction layer is the first layer that encapsulates the user payload data to transport from an SoC system to another SoC system; the resulting packet is called the **Transaction Layer Packet (TLP)** and is formed by adding a header and an **end-to-end cyclic redundancy check (ECRC)** at the end of the packet. It also encapsulates the PCIe requests and responses (known as completions) that don't include user payload data. It can allow up to 4,096 bytes of user data to be transferred from a PCIe device to another PCIe device when the PCIe system configuration allows it. The TLP is passed to the data link layer, which adds a sequence number at the front of the TLP and an error protection code called the **link cyclic redundancy check (LCRC)** at the tail of the TLP, as such forming a **Data Link Layer Packet (DLLP)**. The DLLP is passed on to the physical layer, which adds a Start symbol at the front of the DLLP and an End symbol at the end of the DLLP, as such forming the **Physical Layer Packet (PLP)**.

The following diagram illustrates the format of the PCIe TLP, DLLP, and PLP:

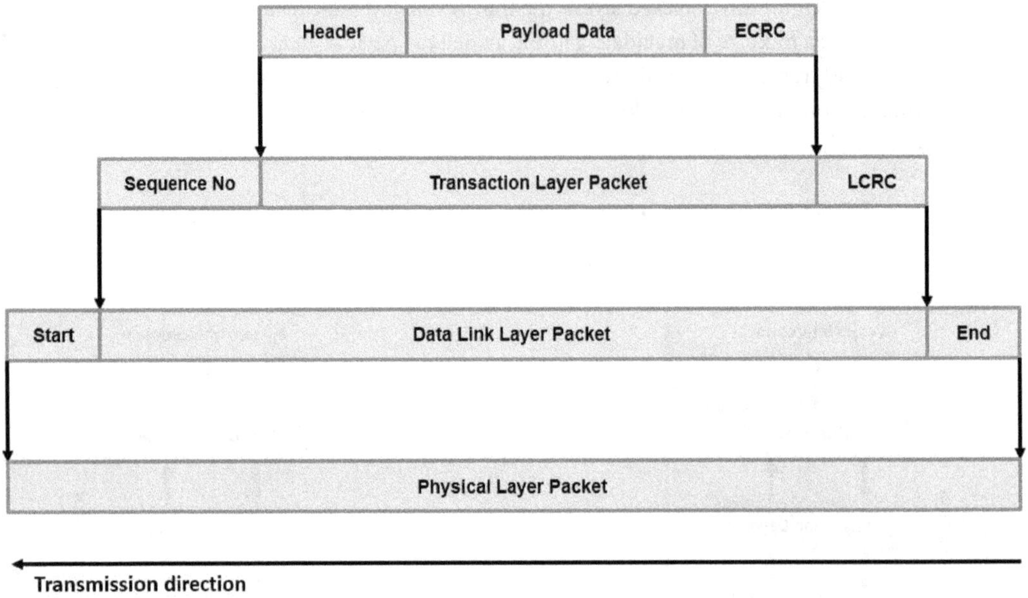

Figure 4.11 – PCIe protocol layer packets

PCIe data link layer

The PCIe data link layer implements the PCIe link management, the TLP ACK/NACK protocol, the link power management, and the flow control information used to exchange the link partner side buffer information. These flow control DLLPs are only 8 bytes in size and are only exchanged between the link partners, that is, are not routable. For further details on the DLLPs, please consult the PCIe specification at https://pcisig.com/specifications.

PCIe transaction layer

The PCIe transaction layer protocol defines four types of address spaces: the memory address space, the I/O address space, the configuration address space, and messages. The I/O address space is only allowed from PCIe devices toward legacy PCI devices. To distinguish which address space is accessed, the TLP header uses a *Type* field to define the transaction that is specific to the address space being accessed. The following table provides a summary of these commands:

TLP Header Type Field	Nomenclature
Memory Read	MRd
Memory Read Lock	MRdLk

TLP Header Type Field	Nomenclature
Memory Write	MWr
Configuration Type 0 Read	CfgRd0
Configuration Type 0 Write	CfgWr0
Configuration Type 1 Read	CfgRd1
Configuration Type 1 Write	CfgWr1
Message Request without data	Msg
Message Request with data	MsgD
IO Read	IORd
IO Write	IOWr
Completion without data	Cpl
Completion with data	CplD
Completion-Locked	CplLk
Completion with data Locked	CplDLk

Table 4.4 – PCIe TLP header type and transaction target address space

The TLP has a header that defines the transaction characteristics and can be 12 **Double Words** (**DWs**) for a 32-bit system addressing mode at the target or 16 DWs for 64-bit system addressing, the DW being 4 bytes in size. The TLP header includes information such as the transaction type, destination address or ID, transfer size for TLPs carrying data, **Quality of Service** (**QoS**) information, and so on. The TLP header has many fields that are dependent on the *Type* field; the following figure provides an overview of the TLP header format. Please note that DW3 is only present in 16 DW TLP headers where the target addressing mode is 64 bits:

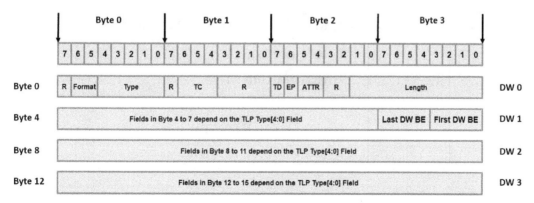

Figure 4.12 – PCIe TLP header format

The following table provides a description of the TLP header fields. Please note that *R* means *reserved* and is set to 0b0:

Field	Size in bits	Description
Format	2	Indicates the header size and whether the TLP transports payload data or not. The encoding is as follows: • 0b00: 3 DW header with no payload data in the TLP • 0b01: 3 DW header with payload data in the TLP • 0b10: 4 DW header with no payload data in the TLP • 0b11: 4 DW header with payload data in the TLP
Type	5	Used with the Format field to indicate the TLP transaction type address space target as described in Table 4.6.
TC	3	Traffic class assigned to this TLP. Along with the PCIe virtual channels, this helps in implementing QoS in PCIe.
TD	1	TLP Digest field present, when set to 0b1; this field indicates whether the TLP includes an ECRC field at the end of the TLP.
EP	1	Poisoned data, when set to 0b1, means that this TLP transaction is allowed to progress, but its included data is invalid.
ATTR	2	Bit 5: Relaxed Ordering, when set to 0b1, means that the PCI-X relaxed ordering is in use; when set to 0b0, then strict ordering among TLPs is to be used. Bit 4: No Snoop, when set to 0b1, means that no host cache coherency is associated with this TLP, and when set to 0b0, means that host coherency is implemented.
LENGTH	10	TLP payload data size in units of DW, from 4 bytes to 4,096 bytes: • 0b00 0000 0001: 1 DW • 0b00 0000 0010: 2 DW • … • 0b11 1111 1111: 1,023 DW • 0b00 0000 0000: 1,024 DW

Field	Size in bits	Description
LAST DW BE	4	Byte enables mapped one-to-one to the last DW of the TLP payload; when a bit is set to 0b1, then the corresponding byte in the last DW is valid, and when set to 0b0, it means that the byte is invalid.
FIRST DW BE	4	Byte enables mapped one-to-one to the first DW of the TLP payload, when a bit is set to 0b1 then the corresponding byte in the first DW is valid, and when set to 0b0, it means that the byte is invalid.

Table 4.5 – PCIe TLP header fields description

The following text provides further details on the TLP header *Type[4:0]* field encoding with *Format[1:0]* defining the transaction type:

TLP	Address Space	Format[1:0]	Type[4:0]
MRd	Memory	0b00: 3DW no payload	0b0 0000
MRd	Memory	0b01: 4DW no payload	0b0 0000
MRdLk	Memory space	0b00: 3DW no payload	0b0 0001
MRdLk	Memory space	0b01: 4DW no payload	0b0 0001
MWr	Memory space	0b10: 3DW with payload	0b0 0000
MWr	Memory space	0b11: 4DW with payload	0b0 0000
IORd	IO space	0b00: 3DW no payload	0b0 0010
IOWr	IO space	0b10: 3DW with payload	0b0 0010
CfgRd0	Configuration space	0b00: 3DW no payload	0b0 0100
CfgWr0	Configuration Space	0b10: 3DW with payload	0b0 0100
CfgRd1	Configuration Space	0b00: 3DW no payload	0b0 0101
CfgWr1	Configuration Space	0b10: 3DW with payload	0b0 0101
Msg	Message	0b01: 4DW no payload	0b1 0xxx *(see note (1))*
MsgD	Message	0b11: 4DW with payload	0b1 0xxx *(see note (1))*
Cpl	Request dependent	0b00: 3DW no payload	0b0 1010
CplD	Request dependent	0b10: 3DW with payload	0b0 1010
CplLk	Memory space	0b00: 3DW no payload	0b0 1011
CplDLk	Memory space	0b10: 3DW with payload	0b0 1011

Table 4.6 – PCIe TLP transaction type encoding

(1): The Type[2:0] is the message routing subfields used as follows:

- 0b000: Implicit routing to the PCIe RC.
- 0b001: Routed by using the address fields of the TLP header.
- 0b010: Routed by ID (bus, device, and function) assigned by the PCIe RC at system setup.
- 0b011: This is an RC broadcast message.
- 0b100: This is a local message; terminate at the receiver.
- 0b101: Gather and route it to the RC.
- 0b110 and 0b111: Reserved.

From a functional perspective, a PCIe transaction is initiated by a master device within the SoC, usually, a CPU that needs to read data or write data to a peer device connected to it over PCIe. For a data read example from the peer device memory, the requesting device will prepare the TLP header as a data structure in its local memory domain and pass it on either directly (via the PCIe bus slave interface) or via registers to its local PCIe controller. The PCIe controller will then transform the request into a PCIe **memory read** (**MRd**) request that is sent to the peer device over the PCIe link. The peer device will then complete the request via TLP **completion with data** (**CplD**), with data usually in chunks of 64 bytes (or higher, 128 bytes or 256 bytes) according to the system setup by the PCIe RC. For writing to the peer device memory, the local CPU will prepare the TLP header as a data structure in its local memory and uses the PCIe controller-provided mechanisms to trigger the PCIe write operation to the peer device. The write operation is a posted transaction that doesn't require a TLP completion from the target PCIe device back to the requesting PCIe device.

PCIe controller example

The Zynq UltraScale+ MPSOC device contains a PCIe controller that can be configured as a root port or an end point. It is PCIe Gen1 and Gen2 compatible, it can be configured as an x1, x2, and x4 lanes in width. It includes a DMA engine and all the necessary interfacing logic to build a full PCIe EP controller or, when in combination with the **PS** CPU, a PCIe RC controller. The following diagram illustrates this PCIe controller:

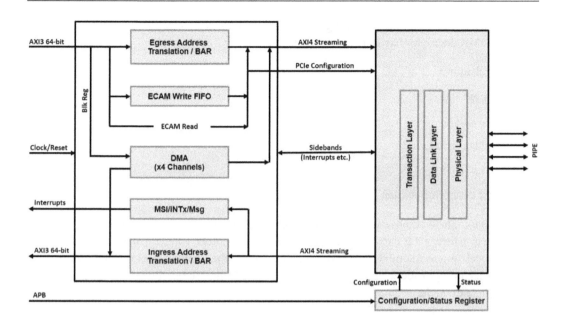

Figure 4.13 – PCIe controller example – the Zynq UltraScale+ MPSOC

> **Information**
>
> For further details on this PCIe controller and all its features, please check section 30 of the *Zynq UltraScale+ MPSOC Technical Reference Manual* at `https://docs.xilinx.com/v/u/en-US/ug1085-zynq-ultrascale-trm`.

PCIe subsystem data exchange protocol example using DMA

When implementing a system-level solution to connect two or more high-speed devices using the PCIe interconnect, there are many other considerations beyond the PCIe-specific requirements that need addressing at the system level. The system software application needs to define a transport protocol that can use the PCIe interface to move data from one device address space domain to the PCIe node partner address space domain. In SoC applications, usually a point-to-point PCIe system topology is used, where one end is the PCIe RC, and the other end is the PCIe EP. It is usual to find that the PCIe EP has an associated DMA engine that is used in the system to move data from/to its address space to/from the PCIe RC address space. The system-level software in the PCIe RC side can define a protocol via which data movement is implemented taking into consideration any security aspects, which is usually a system architecture feature already decided for a given application in its hardware, firmware, and software aspects. In this system-level application example, we consider that the PCIe

EP controller has an integrated DMA engine that defines a data movement using a DMA descriptor, and each DMA descriptor has at least the following entries:

- Local address
- Remote address
- Data size
- Control settings
- Status
- Next descriptor pointer

A field in the control settings entry defines whether this DMA descriptor is a unique DMA operation descriptor, or whether it is part of a linked list of DMA descriptors, where the next DMA descriptor is in the local system memory at the address specified by the next descriptor pointer entry. The following diagram provides a visual description of the DMA descriptors linked list concept defined in this sample example:

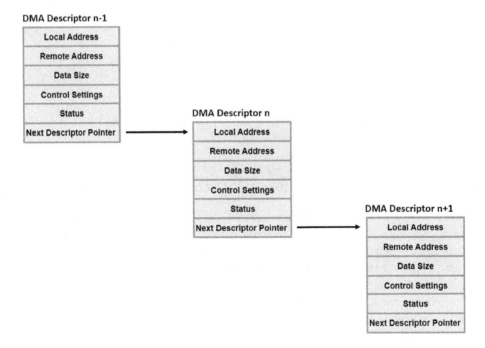

Figure 4.14 – PCIe EP DMA descriptor linked list concept

The next diagram depicts the sample example PCIe subsystem:

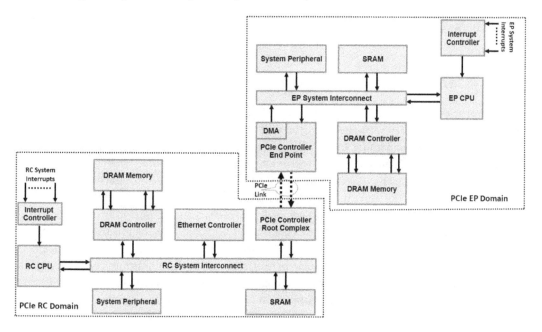

Figure 4.15 – PCIe subsystem sample example

Once the PCIe RC sets up the PCIe subsystem following the topology hierarchy discovery and the EP configuration, the RC CPU can define a data exchange protocol with the EP CPU. The RC CPU defines a memory region in its DRAM memory space and configures the required security access attributes so the EP CPU can access it. The RC CPU may specify, for example, two circular buffers in its DRAM memory space, one for the control data and another for the payload data. It then writes the base address of these circular buffers into a predefined memory address within the PCIe EP domain. These two circular buffers can be used to implement a data transport mechanism between the RC CPU and the EP CPU over the PCIe interface. Assuming that the EP CPU is the data producer, it will then use the PCIe EP DMA engine to write the data from the PCIe EP domain DRAM memory space into the PCIe RC domain DRAM memory space. The DMA descriptors are preset by the RC CPU in its DRAM control data circular buffer, and the pointer to it is already written in a known location by the RC CPU in the EP PCIe domain DRAM memory. The EP CPU sets the PCIe EP controller DMA to read the entries from the remote circular buffer; we assume that the number of entries is preset to N descriptors. Once the PCIe EP controller DMA engine reads the $N \times DMA$ descriptors from the control data circular buffer, it notifies the EP CPU via an interrupt. At this moment, the EP CPU has the necessary preset DMA descriptors in its local DRAM memory space; it may need to add the local pointers to the local source of the data. The EP CPU then sets the DMA engine of the PCIe EP controller using the final version of the $N \times DMA$ descriptors and arms it when the data itself to transfer is ready. Once the DMA operations are finished, the data is transferred over the PCIe link to the RC PCIe domain DRAM memory space using the destination addresses initially set by the RC CPU. Usually, a notification mechanism via a message-based interrupt is sent from the EP CPU to the RC CPU over PCIe. For more details on message-based interrupts (MSI or MSI-X), please consult the PCIe specification at `https://pcisig.com/specifications`.

PCIe system performance considerations

Table 4.3 provides the theoretical PCIe link bandwidth; however, these are the line rates that can be achieved at the physical layer. As already introduced, there are overheads that reduce this bandwidth, such as the TLP header and trailer, the data link layer protocol packets used for the link management, and the physical layer symbols. The PCIe generation physical layer encoding scheme, such as 8b/10b in Gen1 and Gen2, as well as 128b/132b in Gen3, Gen4, and Gen5, also introduces a performance degradation in terms of the data throughput. There are also system-level aspects that affect the PCIe system performance such as the **maximum payload size** (**MPS**), which defines the maximum payload data size that can be moved at any moment by a single TLP transaction, be it a **MRd** or a **memory write** (**MWr**); it has a maximum of 4,096 bytes and usually, it is set by the PCIe RC to a lower value in SoC applications such as 128 bytes or 256 bytes. There is also the **maximum read request size** (**MRRS**), which defines the maximum payload data size that can be requested in a single MRd TLP transaction. Another setting is the **read completion** (**RdC**), which specifies the CplD TLP MPS that can be transferred at a time as a response to an MRd TLP request; the RdC is usually set to 64 or 128 bytes in SoC-based applications using PCIe as a system interconnect. When the RC is lower than the MRRS, the PCIe completer uses as many CplD TLPs as needed to fulfill the MRd request.

> **Information**
>
> For further details and how to theoretically estimate the PCIe read and write effective throughput, you can use the following white paper at `https://docs.xilinx.com/v/u/en-US/wp350`.

Ethernet interconnect

Ethernet-based interconnect is a mature networking protocol, but it keeps evolving in terms of features, physical transport mediums, and connection speeds. It is also one of the simple interfaces that can be used to connect two subsystems together and allow them to share data using a mature connectivity solution. The Ethernet protocol first appeared in 1980, and it was formally standardized under IEEE 802.3 in 1983. We cover Ethernet as an SoC interface because it is included in the Xilinx SoCs covered in this book; it can also allow us to connect two SoCs using the low-level Ethernet protocol as a data transport mechanism, by using a software TCP/IP stack running on an operating system on the Zynq SoC CPUs, or by defining a light data sharing protocol that can use the Ethernet protocol as a transport layer. We will cover communication applications in more detail in the advanced section of this book. In this chapter, we will introduce the Ethernet protocol and the Zynq-7000 SoC Ethernet controller.

Ethernet speeds historical evolution

Since its first appearance in the early 1980s, Ethernet speeds have seen a huge evolution from the initial few MB/s to the currently available hundreds of GB/s speeds. The following table provides a historical overview of the Ethernet theoretical bandwidths:

Ethernet version	Speed	Introduction year
Ethernet (10Base-X)	10 Mb/s	1980
Fast Ethernet (100Base-X)	100 Mb/s	1995
Gigabit Ethernet (1000Base-X)	1 Gb/s	1998
2.5 Gigabit Ethernet (2.5GBase-T)	2.5 Gb/s	2016
5 Gigabit Ethernet (5GBase-T)	5 Gb/s	2016
10 Gigabit Ethernet (10GBase-X)	10 Gb/s	2002 – 2006
25 Gigabit Ethernet (25GBase-X)	25 Gb/s	2016
40 Gigabit Ethernet (40GBase-X) *(1)*	40 Gb/s	2010
50 Gigabit Ethernet (50GBase-X)	50 Gb/s	2016
100 Gigabit Ethernet (100GBase-X) *(2)*	100 Gb/s	2010 – 2018
200 Gigabit Ethernet (100GBase-X)	200 Gb/s	2017
400 Gigabit Ethernet (100GBase-X)	400 Gb/s	2017

Table 4.7 – Ethernet speeds historical evolution

(1) As 4x 10GBase-X

(2) As 10x 10GBase-X or 4x 25GBase-X

Ethernet protocol overview

Ethernet communication uses Ethernet frames at the data link layer of **Open Systems Interconnect (OSI)** to exchange data between a source node and a destination node. For more information on the OSI model, please consult `https://www.iso.org/obp/ui/#iso:std:iso-iec:7498:-1:ed-1:v2:en`.

An Ethernet frame is composed of an Ethernet header, the payload data, and a **frame check sequence (FCS)** field, which is a CRC used to protect the Ethernet frame from errors while transmitted and received between two Ethernet nodes. At the physical layer, Ethernet adds a preamble field and a start frame delimiter at the start of the Ethernet frame and an interpacket gap field at the end of the Ethernet frame. The following diagram illustrates the Ethernet frame, both at the data link layer and the physical layer levels:

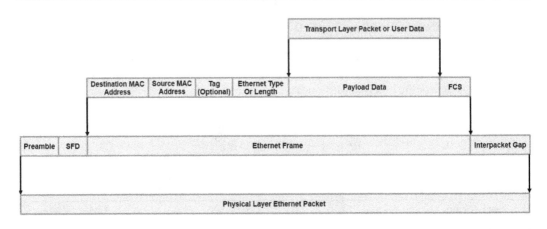

Figure 4.16 – Ethernet frames format

The Ethernet header has many fields, as described in the following table:

Field Name	Size (bytes)	Description
Preamble	7	A pattern of 0b0 and 0b1 used for the receiver clock synchronization.
Start of Frame Delimiter	1	Byte-level synchronization and a new frame delimiter, which uses the value 0xD5.
Destination MAC Address	6	The destination MAC address, which is used for Ethernet frame delivery purposes.
Source MAC Address	6	The source MAC address, which provides the source of the frame.
Tag	4	This is an optional field and when present, indicates the VLAN.
Ethernet Type or Length	2	When this value is 1,500 or smaller, it indicates the length of the payload in bytes. When it is 1,536 or greater, it is the Ethernet Type (EtherType), indicating which protocol is encapsulated under the payload data.
Payload data	42 or 46 – 1,500	Payload data: a minimum of 42 bytes when the frame is tagged, otherwise, a minimum of 46 bytes. The payload could be user data or a frame from a higher-level protocol.

Field Name	Size (bytes)	Description
Frame Check Sequence	4	CRC computed over the entire frame for error protection during transmission.
Interpacket Gap	12	This is the minimum idle time between consecutive packets.

Table 4.8 – Ethernet packet fields description

For further details on the Ethernet protocol, you are encouraged to study the IEEE *802.3* specification available at `https://standards.ieee.org/ieee/8802-3/10556/802.3/7071/`.

Ethernet interface of the Zynq-7000 SoC overview

The Zynq-7000 SoC includes an Ethernet interface capable of operating at three speeds: 10 Mb/s, 100 Mb/s, or 1 Gb/s. This Ethernet MAC is capable of operating in full-duplex mode or in half-duplex mode, depending on the link partner to which it is attached. There are two independent Ethernet MAC controllers in the Zynq-7000 SoC. The Ethernet PHY is provided by a separate external chip to the SoC and can be connected to the SoC via a 24-pin interface called the **Gigabit Media Independent Interface (GMII)**. This requires the use of the **External Multiplexed Input/Output (EMIO)** through the FPGA programmable logic side or via a 12-pin interface called the **Reduced Gigabit Media-Independent Interface (RGMII)** when using the PS's **Multiplexed Input/Output (MIO)**. For further details on the Ethernet PHY GMII and RGMII interfaces and their specifications, you can consult the IEEE standard *802.3-2008, Part 3: Carrier Sense Multiple Access with Collision Detection (CSMA/CD) access method and Physical Layer* specifications at `https://standards.ieee.org/ieee/8802-3/10556/802.3/7071/`.

> **Information**
>
> For further details on the EMIO and MIO multiplexed I/O that can be connected to the Zynq-7000 SoC PS, please consult the *Zynq-7000 SoC Technical Reference Manual* at `https://docs.xilinx.com/v/u/en-US/ug585-Zynq-7000-TRM`.

The Ethernet controller has an integrated DMA engine that interfaces to the SoC system memory through an AHB master interface. Additionally, it has an APB slave interface through which the PS CPU can configure the Ethernet MAC and check its operating status at runtime.

The following diagram provides a micro-architectural-level view of the Zynq-7000 SoC Ethernet controller:

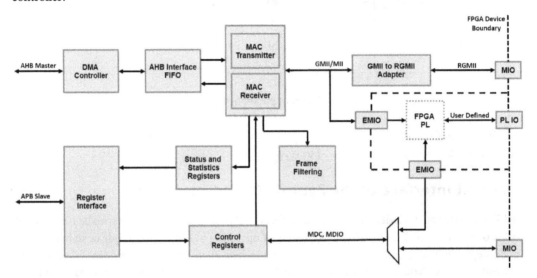

Figure 4.17 – Zynq-7000 SoC Ethernet controller microarchitecture

The DMA engine uses a list of buffer descriptors that software running on the PS CPU prepares for the DMA engine to use when transmit or receive Ethernet operations are started. There are two channels within the DMA engine, one channel for receive and another channel for transmit operations. Each channel has its own list of DMA descriptors. The DMA descriptors are continuous in the PS CPU memory space with the option to wrap when the list reaches its end. Each list is indicated by the PS CPU to the DMA channel engine via a programmable register. Each descriptor has an ownership bit that allows descriptor recycling, meaning that for the receive path, when the software consumes the buffer pointed to by the descriptor, the CPU flips the ownership field, indicating to the DMA engine that it has consumed the buffer pointed to by this descriptor and that now, the DMA engine is allowed to reuse it; if not, then the DMA engine will not. For the transmit direction, software should reuse the DMA descriptor only when it has been transmitted and the ownership bit has been reset accordingly by the DMA transmit engine. For more details on the DMA descriptors and the Ethernet controllers included in the Zynq-7000 SoC, please consult section 16 of the *Zynq-7000 SoC Technical Reference Manual* at https://docs.xilinx.com/v/u/en-US/ug585-Zynq-7000-TRM.

Introduction to the Gen-Z protocol

This section introduces the Gen-Z standard, an emerging memory-semantic protocol that aims to standardize memory and storage accesses and share them between CPUs, **graphic processing units (GPUs)**, and accelerators in heterogeneous systems. This initiative is led by an industry consortium formed 5 years ago by 12 companies; for more information about the Gen-Z Consortium and its forming member companies, please check https://genzconsortium.org/.

Gen-Z itself is now in the process of being integrated into the **Compute Express Link (CXL)** standard developed by the CXL Consortium; for more details on the CXL protocol, please check `https://www.computeexpresslink.org/about-cxl`.

Gen-Z protocol architectural features

The Gen-Z protocol decouples the load-and-store operations from the local controller and interposes the Gen-Z as a fabric in between them. The following diagram illustrates the main architectural concept of the Gen-Z fabric:

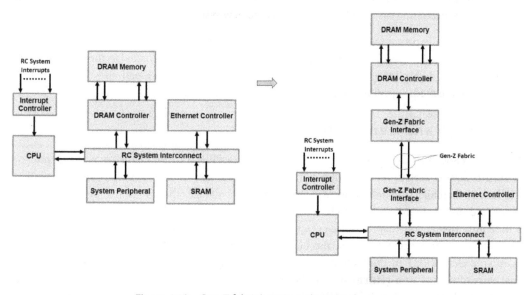

Figure 4.18 – Gen-Z fabric's main architectural concept

With this decoupling, the CPU to memory byte-addressable load-and-store model can be extended to the entire system, where the load-and-store requests are transported first over the fabric, then addressed or translated locally within the media controller. This is possible in modern SoCs where the speed and throughput of the interconnect and the media controllers are high enough to make sharing data using this approach in modern heterogeneous system architectures a potential optimal solution in terms of silicon cost and power consumption. The Gen-Z fabric supports asymmetric data paths, where the read path can be wider than the write path, and vice versa. Access can be local and distributed over memory locally within the system or relatively remotely over a network when access latency makes sense and networked sharing is the optimal option. Gen-Z supports data rates at 16, 25, 32, 56, and 112 GT/s. The Gen-Z protocol abstracts the media type from the fabric as such making access to DRAM, Flash, and storage class memory load-and-store operations uniform providing more flexibility and wider choices for modern SoC designs. The media controllers become abstracted and get located within the local media domain than being interlocked with the CPU or any other compute node. This approach also allows load-and-store operations to be switchable beyond the SoC local interconnect, and the extra added latency will be dependent on the underlying physical layer upon which the Gen-Z will be transported.

The Gen-Z protocol uses memory semantic operations defined using OpCodes and OpClasses that optimize the data transfers and minimize the SoC interconnect utilization as such reducing the overall system power requirements for the same performance in legacy systems. For example, data from the SoC DRAM can be moved to Flash memory in one load and one store operation using the corresponding OpCode/OpClass Gen-Z commands. This also makes the future generation SoCs more scalable and modulable as compute and storage resources can be added without re-architecting the SoC if the need arises in the future, now that compute and storage elements have become plug-and-play to a certain extent and are becoming easily tailored to the specific new system needs.

Gen-Z additionally promotes system low latency by eliminating the need for software stack layers to perform the media-specific translations. This is operationally now offloaded to the media-specific controller sitting behind the Gen-Z fabric that only requires a common load-and-store command regardless of the media-specific access patterns and access latencies.

SoC design and Gen-Z fabric

Building an SoC that can benefit from the Gen-Z features while it is still maturing as a new industry standard requires careful consideration, but it is still a good idea to continue following its evolution. Modern SoC architecture definitions will be faced with challenges that the Gen-Z protocol can help solve; in doing so, there is a need to find suitable Gen-Z controller **intellectual property** (**IP**). This will require media controllers, fabric switches, and Gen-Z interfaces; these are most likely already available from a few vendors and can also be considered as a design activity worth engaging in. There is also the Gen-Z PHY mediums to choose from, and the existing first system implementations will make use of mature **non-return-to-zero** (**NRZ**) transceivers with their achievable high speeds and very good reliability, such as PCIe PHY at 16 and 32 GT/s, and the Ethernet PHYs at 25 GT/s. There is also the need for verification IPs and FPGA-based prototyping solutions, which are also starting to be announced by members of the Gen-Z Consortium.

CCIX protocol and off-chip data coherency

The CCIX protocol is a chip-to-chip data coherent protocol that enables one or more interconnected nodes to share data in a cache-coherent manner without the need for any software cache maintenance operations. It has been developed by the CCIX Consortium to solve the need for the emerging heterogeneous systems architecture driven by the immense data processing needs of modern SoCs in many emerging applications such as big data, autonomous driving, IoT, machine learning, and artificial intelligence. The CCIX Consortium includes 6 industry promotors, 14 contributing companies, and 16 industry adopters at present. For more information about the CCIX Consortium, please check `https://www.ccixconsortium.com/`.

CCIX protocol architectural features

The CCIX Consortium has been focusing on building an **Instruction Set Architecture (ISA)** independent data coherency protocol that can be used to connect accelerators such as GPUs and hardware accelerators (built within FPGAs or in custom-made ASICs) to SoCs using different CPU architectures. CCIX also supports inter-chip virtual addressing modes so data sharing can pass pointers from application-level layers without the need to linearize the data in memory by using in-place DMA operations that add latency and consume more energy. The architectural concept of the CCIX protocol is illustrated in the following diagram:

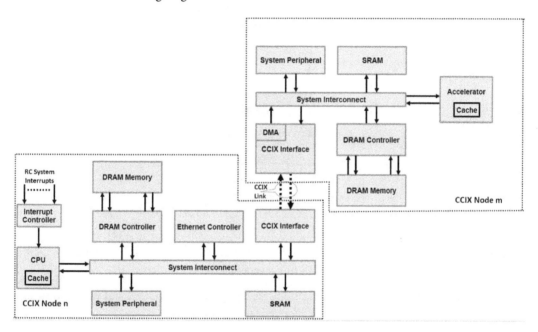

Figure 4.19 – CCIX protocol architectural concept

The CCIX protocol has a layered architecture that sits on top of the PCIe data link layer. The CCIX protocol doesn't use the PCIe transaction layer, but rather defines three more stacked components, as illustrated in the following diagram:

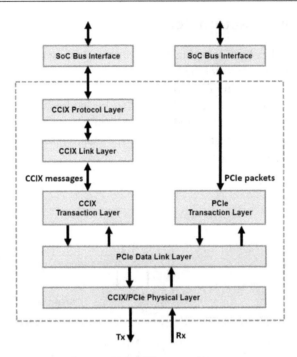

Figure 4.20 – CCIX protocol layers

The CCIX protocol specification is divided into two main sections. The first is the CCIX protocol itself, which defines the inter-chip cache-coherent protocol, the messaging layer, and the flow control mechanisms; the second specifies the CCIX link layers that transport them.

The CCIX link layers or transport specifications are composed of the CCIX and PCIe transaction layers, the PCIe data link layer, and the CCIX physical layer.

CCIX protocol layer

The cache coherency protocol is implemented by the CCIX protocol layer, and it is the top portion of the CCIX standard responsible for MRd and MWr. This layer provides a mapping for the on-chip cache coherency protocol such as the ARM **Advanced Microcontroller Bus Architecture (AMBA) Cache-Coherent Hub (CHI)**. The cache states defined by this layer allow the cache coherency hardware to determine whether a given cache line in a CPU or any accelerator hardware is clean or dirty, and unique or shared. SoC cache coherency has been covered by *Chapter 3, Basic and Advanced On-Chip Busses and Interconnects,* of this book. For more information on the ARM AMBA CHI protocol, you are encouraged to consult the AMBA specification at `https://developer.arm.com/architectures/system-architectures/amba/specifications`.

CCIX link layer

This layer is below the CCIX protocol layer, and it formats the CCIX commands for the transport layer in use. In the current revision of the CCIX specification (revision 1.1), the transport layer is based on PCIe.

CCIX and PCIe transaction layer

The CCIX traffic is assigned its own **virtual channel** (**VC**) that differentiates it from the PCIe traffic, which is assigned to other VCs. This approach allows the CCIX traffic and PCIe traffic from the upper layers to share the same PCIe controller and the same PCIe link. The CCIX protocol can use the standard PCIe packet formats, but it can also map to its own optimized packet formats, which eliminates many unneeded fields in the PCIe packet header.

PCIe data link layer

The CCIX protocol uses the PCIe data link layer as it is. As already covered in this chapter, the PCIe data link layer performs the PCIe/CCIX link management operations as well as transporting the PCIe/ CCIX TLPs by adding an error checking mechanism using the **LCRC**.

CCIX/PCIe physical layer

The CCIX physical layer supports all the PCIe speeds up to Gen5, that is, 2.5 GT/s, 5 GT/s, 8 GT/s, 16 GT/s, and 32 GT/s. CCIX also adds support for 20 GT/s and 25 GT/s.

> **Information**
>
> For a detailed introduction to the CCIX protocol architecture, you are encouraged to study the white paper at https://www.ccixconsortium.com/wp-content/uploads/2019/11/ CCIX-White-Paper-Rev111219.pdf.

Summary

This chapter began by giving an overview of the buses and interconnects used off-chip to connect an FPGA-based SoC to other high-speed devices on the electronics board. It looked at two legacy system buses: the SPI and the I2C. It provided a detailed overview of the PCIe interconnect, its stacked protocol layers, and systems consideration aspects in terms of performance and application-level usage. It then introduced the Ethernet protocol and its suitability to interconnect SoCs on a system or network them using higher software middleware layers such as the TCP/IP protocol. The chapter also looked at the emerging Gen-Z protocol and its concept of using a uniform semantic to access memory over a fabric and remove the CPU to memory and to any storage media interlock. We also covered the emerging CCIX cache-coherent interconnect protocol and the concept of extending data coherency off-chip by adding the necessary protocol layers to manage it in hardware and then transporting it around the system beyond the SoC boundaries over PCIe, for example. The chapter covered the tight link between the PCIe protocol layers and the CCIX protocol layers and how they can be merged and use the same data link and physical layers by exploiting the PCIe protocol QoS and its VCs.

In the next chapter, we will continue introducing the main SoC interfaces required to manage and use internal and external system resources such as internal and external memory and storage. We will also cover most of the communication interfaces available within the Xilinx Zynq-7000 SoC.

Questions

Answer the following questions for rest your knowledge of this chapter:

1. Describe the main features of the SPI bus protocol.

2. How can the I2C bus protocol interconnect a system with many masters and several slaves?

3. What are the layers of the PCIe protocol stack?

4. What are the supported speeds and widths of the PCIe link?

5. What are the main components in a PCIe RC? How many PCIe RCs can a PCIe system have?

6. What are the PCIe maximum payload size, maximum read request size, and read completion? What is the relationship between them?

7. How can the read completion affect the read performance over PCIe at a system level?

8. Describe how a payload size of 9,000 bytes can be written from DRAM memory in a PCIe end point node, using its DMA engine, into DRAM memory within a PCIe RC node.

9. What are the main architectural features of the Ethernet protocol?

10. How can Ethernet interconnect two SoCs and move data between them without using any higher-level middleware or TCP/IP stacks?

11. What are the main purposes of the Gen-Z protocol? Who owns it?

12. How can the Gen-Z protocol improve the system performance of a memory-centric application? Provide an example with a simple quantitative analysis.

13. Describe the CCIX protocol and its main architectural features.

14. Who owns the CCIX protocol? When can it be used?

15. What is the performance improvement that the CCIX protocol can bring to a system with multiple CPUs using different ISAs, and multiple accelerators using internal caches? Provide a quantitative analysis on a simple example you can think of.

16. Which PCIe protocol layers are common between the PCIe and CCIX protocols? How can they share the same medium although they are different protocols serving different application layers?

5

Basic and Advanced SoC Interfaces

In this chapter, we will define an SoC interface for a given function. Then, we will look at the different memory interfaces for on-chip and off-chip memories, their topologies and architectural features, and the criteria for choosing a given interface (or a combination of many). We will cover the fundamentals of the processor caches and the **memory management unit (MMU)** while focusing on the Cortex-A9 implementation. Finally, we will explore the main data storage interfaces.

In this chapter, we're going to cover the following main topics:

- Interface definition by function
- Processor cache fundamentals
- Processor MMU overview
- Memory and storage interface topology

Interface definition by function

As a generic definition, an interface in an SoC is a hardware block that acts as a proxy through which transactions are converted from one domain into another. The domain itself depends on the type of interface in question and what transformation it performs. We can distinguish many functions in an SoC topology, some of which are as follows:

- **Type 1**: A peripheral that performs a dedicated task in the hardware within the same chip, such as an interrupt controller, a system timer, or a central **direct memory access (DMA)** engine.
- **Type 2**: A peripheral that performs a dedicated task in the hardware in another chip external to the SoC, such as a temperature sensor, a **radio-frequency integrated circuit (RFIC)**, or an **analog-to-digital converter (ADC)**.

- **Type 3**: A peripheral that acts as an **input/output (I/O)** device such as a **general-purpose input/output (GPIO)** device or a display controller.

- **Type 4**: A peripheral that performs a low-speed communication task, such as a **universal asynchronous receiver/transmitter (UART)**.

- **Type 5**: A peripheral that performs a high-speed communication task, such as an Ethernet controller, a Bluetooth controller, or a **wireless local area network (WLAN)** controller.

- **Type 6**: A bridge from one **network-on-chip (NoC)** to another or interconnect within the same SoC, such as an AXI to APB bridge, a high-speed NoC to a low-speed and low-power peripheral interconnect, or the **processing subsystem (PS)** to the **programmable logic (PL)** bridge within the Zynq-7000 SoC, as introduced in *Chapter 1*.

- **Type 7**: A bridge between the SoC chip and another high-speed **integrated circuit (IC)** residing within the same electronics board, such as a **Peripheral Component Interface Express (PCIe)**.

- **Type 8**: A bridge between the SoC chip and another low-speed IC residing within the same electronics board such as a **serial peripheral interface (SPI)** or an **inter-integrated circuit (I2C)** bus controller.

- **Type 9**: A plug-and-play controller such as a **universal serial bus (USB)** controller.

- **Type 10**: A form of local storage within the SoC chip, such as **read-only memory (ROM)**, **static random Access memory (SRAM)**, or **one-time programmable (OTP)** arrays.

- **Type 11**: A form of external storage to the SoC chip, such as **Dual Data Rate Synchronous Dynamic Random Access Memory (DDR SDRAM)**, **Low-Power Dual Data Rate Synchronous Dynamic Random Access Memory (LPDDR SDRAM)**, or flash memory.

- **Type 12**: A hardware accelerator implemented within the SoC chip, such as a security accelerator engine or a mathematical computation engine.

The common factor between all these functions is that they all connect to the SoC internal NoC or interconnect from one side, and either connect to the chip's internal resources or external I/O pads propagating their reach off-chip. The following diagram depicts the interface by function concept:

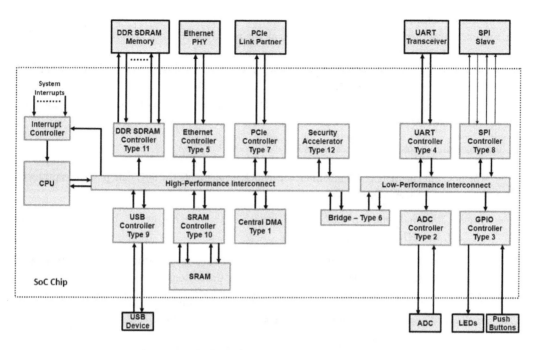

Figure 5.1 – SoC interfaces and their type classification

SoC interface characteristics

As mentioned previously, the interface bridges two domains, and as such, it is required to implement all the services needed to make this bridging possible. In the transfer function fulfilled in this bridging, usually, the clocking is different, either increasing or decreasing, as well as being asynchronous. This bridging can be unidirectional or bidirectional with some or no dependencies between the two data transfer directions. Often, the processor needs to set the operating mode of the function, check the status of operations, and intervene when necessary to either take on follow-up actions or reset the functionality of the device in case of errors. These are provided as a set of configuration and control registers that software programs use when powering up the SoC or when a mode change in how the function operates is desired. The data transfer process, as covered in *Chapter 3, Basic and Advanced On-Chip Busses and Interconnects*, can either be performed by the processor itself from the function's temporary storage buffers or automatically triggered when data is available and then transferred via an integrated DMA engine within the function itself. Usually, domain crossing is done through FIFOs, which also act as temporary storage while certain conditions are met to transfer the data to its destination. The function also uses a set of registers to log operational information of interest to the processor, gather runtime and functional statistics to analyze the behavior, and estimate the function's performance. The function also provides notifications to the processor via a set of interrupts or a single interrupt acting as a proxy for many lower-level interrupts. There is usually an array of registers for logging all the necessary information that the processor can read to decide on further actions and reset operating conditions, when necessary, in case of operational errors, for example. The function has a reset control mechanism that implements all the reset functionalities required by the function itself, such as the system reset, the interfaces reset, the external device controller reset, and so on, as defined by the function microarchitecture. When we look at all these operational details that a function acting as an interface requires, except for Type 6, which acts as an internal SoC bridge from one NoC to another or interconnect, we notice that there is a common set of services that an interface implements (some or all) to fulfill its transfer function.

In summary, these interface services are as follows:

- The SoC internal data plane bus interface, which implements the SoC data plane bus protocol, such as an AXI, ACE, or OCP

- The SoC internal control plane bus interface, which implements the SoC bus protocol, such as an APB, an AHB Lite, or an AXI Stream

- The SoC interconnect clock domain to the function's hardware internal clock domain crossing in the write or transmit direction (as seen from the processor's perspective)

- The function hardware's internal clock domain to the SoC interconnect clock domain crossing in the read or receive direction (as seen from the processor's perspective)

- The transmit/write FIFO or buffer within the function hardware block

- The receive/read FIFO or buffer within the function hardware block

- The DMA write channel within the function hardware block

- The DMA read channel within the function hardware block

- The interface's main transfer function that's implemented within the function's hardware

- The function's configuration, control, and status registers file

- The function's interrupt mechanism implementation with all the registers required to enable, mask, and trigger the function's single or multi-interrupts mechanism

- The function and interfacing logic system reset control

The following diagram depicts the function services concept in an SoC interface:

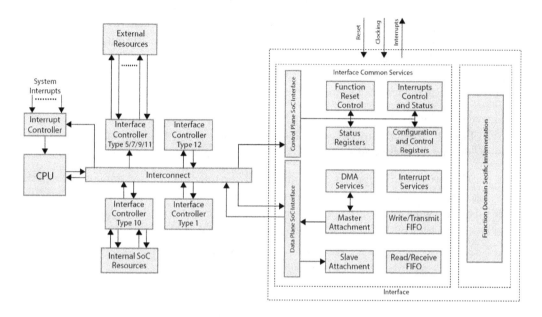

Figure 5.2 – Common interface services

The aforementioned function services are not necessarily present in all SoC interfaces as their range of complexity is wide, and the interface tends to implement the services required to perform the desired transfer functions. When implementing our custom interface in the advanced section of this book, we will start the microarchitecture design of the interface controller from a known set of services and design the interface using the services that provide the desired interface transfer functions.

SoC interface quantitative considerations

When designing, or sometimes even when configuring, an SoC interface provided by a third-party vendor as a configurable IP that can be tailored to a user SoC application, other quantitative considerations will dictate whether certain services are needed, such as using an integrated DMA engine or just relying on the SoC central DMA engine when it has one, using a FIFO and which size to implement, and which frequency to clock the core function at, for example. In these cases, knowledge of the SoC system's performance requirements is a necessity for defining these elements. The SoC's overall architecture should have defined these performance parameters or at least provided the system requirements from which to derive them.

When looking at the overall SoC interface performance, the main data bus and the type of runtime data transfer supported by the interface, such as the burst size, the **issuing capabilities (IC)** of the bus interfaces, the frequency at which it is running, and any interface internal buffering capabilities, help in computing the overall internal SoC side performance of the interface. The bus protocol transfer metrics are provided by the bus standards specification. The ARM AMBA bus collection and the OCP bus protocol were introduced in *Chapter 3, Basic and Advanced On-Chip Busses and Interconnects*, where some of the parameters affecting their performance were covered.

The interface controller's domain-specific performance figures are usually set and predefined by the protocol (if any) to which the interface is performing the transfer function – for example, 1 Gbps for a Gigabit Ethernet controller and 3.4 Mbps in the **high-speed mode** of an SPI controller.

Given that the bus protocol supports data transfers via the interface, the domain protocol-specific metrics for which the controller is interfacing, and the interface internal function buffering capabilities, we can usually compute the interface performance figures and make decisions regarding the interface function configuration or choose a particular SoC implementation.

Processor cache fundamentals

Most processors used in modern SoCs include a **Level 1 Instruction Cache (L1I$)** and a **Level 1 Data Cache (L1D$)**. Some processors also include an **L2 Shared Cache (L2S$)**, which is used for both processor instructions and data per processor core, or like in the ARM Cortex-A9 cluster where it is used for both instructions and data, and shared between all the cores. In some modern processors in a multi-core processor cluster, there is also an **L3 Common Cache (L3C$)** between all the processors in the cluster. The caches are used to shorten the latency of the processor's access to instructions and data while executing the software as they are implemented using SRAMs and running at the processor clock frequency, which is relatively higher than the remaining logic surrounding the processor in the SoC. Also, external memory access latency is usually many orders of magnitude higher than the access time to the internal SRAM implementing the cache within the processor core. This reduces the bus accesses performed by the processor to frequently running code and accessed data of the software, therefore reducing both the bottlenecks on the SoC interconnect and the shared memory hosting the software image and thus optimizing the overall SoC power consumption. The power is reduced by shortening the data movement on the SoC busses and avoiding unnecessary access to external memory. Access to external memory uses external pads which consume more power in comparison to accessing internal SRAM or ROM memory. In this section, we will focus on the processor cache organization in general and look at the ARMv7 architecture that's used in the Cortex-A9 processor, which is implemented within the PS of the Zynq-7000 SoC earlier in this book. The following diagram depicts the hierarchical view of the processor caches in the Cortex-A9 processor cluster:

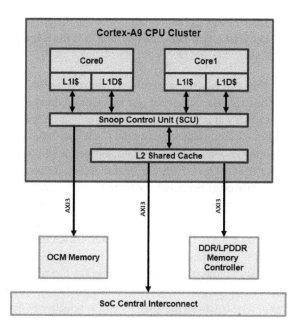

Figure 5.3 – Cortex-A9 processor cache hierarchy

The **Snoop Control Unit (SCU)** is responsible for maintaining the cache coherency between the different caches and the external memory by implementing a cache coherency protocol called MESI, which was introduced in *Chapter 3, Basic and Advanced On-Chip Busses and Interconnects.*

Processor cache organization

When the processor needs access to data or instructions held in the external memory, the address of that data or instruction is looked up in the first-level cache – that is, the L1D$ for data or the L1I$ for the instructions. If there is a hit – that is, this address is held in the L1D$ or L1I$ – then the data or the instruction is returned from the cache. If there is a miss, then the next-level cache – in this case, the L2S$ – is looked up and the data or instruction is returned from the L2S$ in the case of a cache hit; otherwise, the request is forwarded to the physical memory.

The memory content is loaded into the cache in cache line units. For the Cortex-A9, a cache line is 32 bytes in size. Cache memory sizes are limited to avoid increasing the SoC cost and slowing down the processor clock speed. Due to this, managing the use of the caches and placement of the code and data in the memory of the SoC software is a key element in the SoC system performance metric, which we will study in *Part 3* of this book. When observing the software execution on an SoC processor, two aspects are noted:

- The code's locality, which reflects the instructions and data that are executed at a specific period. These tend to be close to each other (in terms of addresses – that is, the physical location in memory, such as when executing loops).

- The other aspect is that software execution tends to reuse the same addresses over time (such as when calling the same function many times), which is the temporal locality aspect of the software's code and data.

Using caches in an SoC processor has the potential to improve the system software's execution performance by reducing the external memory access latency and decreasing the energy cost of the software execution. However, this will introduce some degree of non-determinism that the application should be aware of to define a correctly behaving system. Caches are smaller in size than the external memory and are usually not big enough to hold all of the software application and its associated data. So, when code is fetched for the first time from external memory, its execution time acts as though the cache isn't present; it is only on subsequent access to the same code and data while still in the cache that the system will benefit from the reduced latency and increased performance. This induced non-uniform latency can be a concern for hard real-time systems that require a deterministic response time, regardless of the use of a cache. In these systems, deadlines should be estimated and bound as if the cache is non-existent by placing the code and data in non-cacheable regions or **tightly coupled memories** (**TCM**) for processor architectures that have one. The Cortex-A9 doesn't have this, but it does have an **on-chip memory** (**OCM**) provision in the PS that can be used as a non-cacheable memory region to hold critical data and instructions for software that is sensitive to determinism, for example.

Cache topology and terminology

The following diagram shows how the ARM cache is organized and the associated terminology necessary to understand the operating behavior of the caching mechanism in an ARMv7-A architecture:

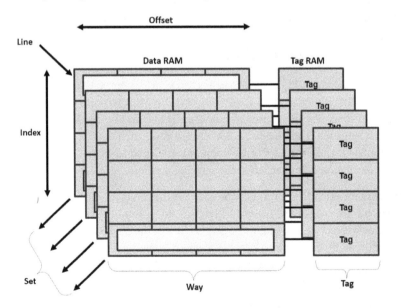

Figure 5.4 – ARMv7-A cache organization

The following table provides a summary of the terms used in the ARMv7 architecture cache's organization:

Term	Definition
Cache	A storage location for the processor's data and instructions.
Data RAM	This stores the address content of either the data or the instructions.
Tag RAM	This stores the address information.
Line	This is the smallest data or instructions unit from the system memory that the cache operates on.
Tag	This is part of the cache line address in the system memory that's stored in the cache.
Index	This is part of the cache line address in the system memory and can be found in the cache.
Offset	This is part of the cache line address in the system memory and via which a specific word can be found in the cache line.
Way	This is an equal-size logical grouping of multiple cache lines.
Set	This is a logical grouping of cache ways looked up simultaneously for a specific cache line. They have the same index values.

Table 5.1 – Cache organization terminology

Directly mapped cache

Since the cache is smaller in size than the main SoC memory, a mapping between the cache and the main memory needs to be implemented to define where the specific memory region of a cache line size can be stored in the cache. We distinguish caches that are directly mapped and others that are set-associative. In a directly mapped cache, a memory address can only be held in a single cache location, as shown in the following diagram:

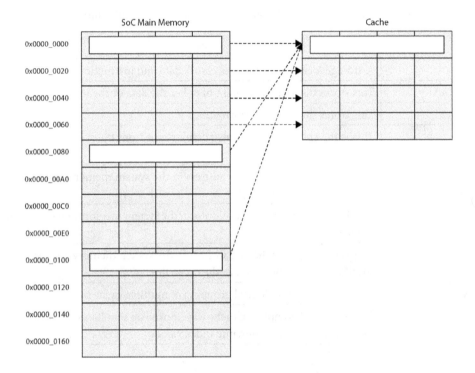

Figure 5.5 – Directly mapped cache

In the preceding diagram, the index bits (address[5:4]) are used to read the tag values stored in the cache. If there is a match and the address line is valid, then there is a hit. The line (address[3:2]) and the offset (address[1:0]) are used to locate the relevant data value within the cache line. The value of the cache line address to store is the remaining 26 bits (address[31:6]), which will be looked up in the tag memory to find out whether the memory location is cached when the processor needs to access it. The following diagram illustrates the cache line address utilization for a directly mapped cache:

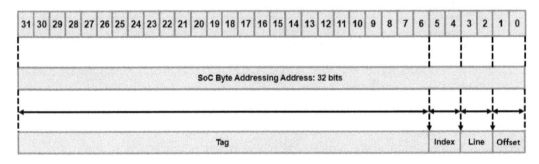

Figure 5.6 – Directly mapped cache line address format

Set-associative cache

As shown in the preceding diagram, memory addresses will be competing for the same cache line location in the cache and causing cache trashing and increasing traffic between the cache and the system memory, which results in higher power consumption and lower system performance. To circumvent this, we need to have multiple ways to host the memory address in the cache. This can be achieved with set-associativity. A set-associative cache increases the number of locations where a memory address can be stored in the cache, thus providing more chances to keep the cache line replacement low. The following diagram illustrates the set-associative cache approach:

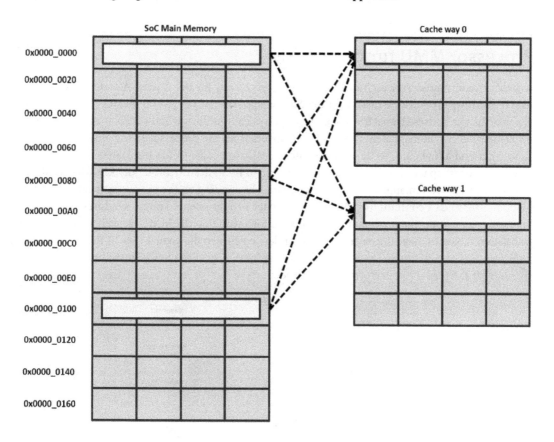

Figure 5.7 – A two-way set-associative cache organization

In this cache organization, the capacity of the cache is split into many ways. A memory location can then be mapped to a way and not to a specific line like in the directly mapped cache introduced in the previous section. The line is still identified using the index field of the address in each way of the set. Consequently, cache lines with the same index values belong to the same set. For a cache hit, we must look at every tag in the set. In the ARM cache implementations of L1I$ and L1D$, two-way and four-way set-associative caches are usually used. In the L2S$, and because the L2S$ is usually bigger by at least an order of magnitude than the L1 caches, the number of ways is higher, and usually 8- or 16-way set-associative caches are common.

For more information regarding the ARMv7 architecture cache's implementation, please read *Section 8* of *Caches of the ARM Cortex-A Series Programmer's Guide* at https://developer.arm.com/documentation/den0013/latest.

ARM Cortex-A9 Technical Reference Manual also contains some useful architectural-specific details on caches for the Cortex-A9: https://developer.arm.com/documentation/ddi0388/latest.

Processor MMU fundamentals

Another modern processor architectural feature, specifically for the Cortex-A series CPUs, is the MMU. It allows the software tasks running on the processor to use virtual memory addressing as it performs the address translation back to the physical memory when access is needed. It also implements access management in the address spaces from the processor to the physical memory to prevent and allow read, write, and execute access from tasks to the translated regions of the physical address space. The MMU allows software to be written without any knowledge of the physical address space mapping or whatever else may be running on the processor. It can also make mappings between a continuous virtual address space to a fragmented physical address space. The firmware running on the processor is responsible for setting up the MMU, as well as the address translation tables, to make the desired correspondence between the virtual view of the SoC address map seen by the software to the actual physical address map of the SoC. The following diagram shows what the MMU can perform in this respect:

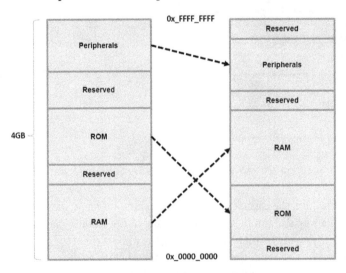

Figure 5.8 – MMU address translation operation

The virtual address space is on the left side of the preceding diagram, whereas the physical address map is on the right side.

The MMU can be disabled (it is at boot time by default) and in this case, the virtual address map is mapped directly to the physical address map with all types of access enabled. The MMU performs the address translation by using page tables (called **translation tables** in ARM architecture terminology) that are created by software in the SoC's physical memory.

The addresses generated by a processor are virtual addresses that are then translated into physical addresses as they cross the MMU when enabled. If we think of the address space as a set of contiguous regions with different sizes, these regions can be mapped anywhere in the address map using several of the most significant bits. Effectively, this is the basis of the translation job performed by the MMU. Each region should also have some attributes that define the permissions that are checked as the translation is performed. If these permissions are not allowed or the region is not mapped, then an exception fault is generated by the MMU. *Translation table walking* refers to the process of reading the translation tables from memory by the MMU hardware during the translation process.

The **translation look-aside buffer** (**TLB**) is used as a caching mechanism by the MMU hardware for the translation tables. So, rather than fetching from memory at every translation operation, it keeps the recently used translation tables in this buffer (or cache), making translating the address from virtual memory to physical as fast as possible. When the translation table is accessed from the TLB, this is called a TLB hit.

The contiguous memory regions in the mapping job of the MMU are called translation pages. The smaller a page is, the better it is to efficiently use the memory space when allocation is done for different software processes that shouldn't share memory, for example. However, this introduces a higher number of translation tables and therefore, potentially inefficient use of the TLB. For example, a software process requires 15 KB of memory. Here, assigning 4 x 4 KB pages is fine since only 1 KB is wasted, but for a process that requires 128 KB, if we assign a page that's 1 MB, then a lot of memory space is wasted. However, with larger page sizes, TLB entries are more likely to maximize the hit rate, thus avoiding external memory for the translation table walking, which decreases performance. 16 MB *supersections* can be used for larger memory region allocations to processes where a detailed and restrictive mapping is required. To come up with an efficient translation scheme for a given SoC software, specifically when not using an OS, you may need to experiment with a few translation schemes using different page sizes to find the best compromise.

The MMU address translation process in a Cortex-A9 processor can involve many steps and uses multiple translation tables in its implementation.

The process of the address translation and how many steps are included in it for the ARM Cortex-A9 processor are described in detail in *Section 9* of the *ARM Cortex-A Series Programmer's Guide* at `https://developer.arm.com/documentation/den0013/latest`.

The *ARM Cortex-A9 Technical Reference Manual* also contains some useful architectural-specific details on the MMU and its hardware implementation for the Cortex-A9: `https://developer.arm.com/documentation/ddi0388/latest`.

Memory and storage interface topology

In this section, we will focus on the memory and storage controllers available in the PS of the Zynq-7000 SoC since we will need to configure them for *Part 2* of this book. We will examine their microarchitectural features, their main configuration parameters, and how they affect the system performance aspects. In this section, we will cover the following Zynq-7000 SoC interfaces:

- DDR memory controller
- Static memory controller
- On-chip memory controller

Let's get started.

DDR memory controller

The DDR memory controller that's integrated within the PS of the Zynq-7000 SoC can be used to interface to DDR2, DDR3, DDR3L, and LPDDR2 type SDRAM memories. It is a multiport controller from the SoC side, and its bandwidth can be simultaneously shared between its four ports for both read and write transactions. It has four AXI3 internal ports that are full duplex. A port is connected directly to the L2 cache of the Cortex-A9, allowing it to service the processor cores and making the DDR memory cache coherent via the ACP of the Cortex-A9 with any external master to the PS. Another port is shared between all the internal masters within the PS via the central interconnect, and the two remaining ports are dedicated to the SoC programmable logic through two AXI3 64-bit full duplex interfaces. The DDR controller is formed of three main microarchitectural blocks:

- **DDRI**, which interfaces to the SoC interconnect via the 64-bit four AXI3 interfaces. It provides stage-1 arbitration access and buffers the command and data using the implemented internal FIFOs.

- **DDRC**, which is the core of the controller with the necessary logic to perform the per-port read/write stage-2 arbitration and the transaction scheduling in an optimal way (stage-3). This maximizes the bandwidth utilization of the attached external DDR DRAM memory while also minimizing latency for urgent transactions.

- **DDRP**, which is the PHY interface that translates the DDRC selected commands into DDR DRAM-specific transaction signals that obey its timing sequencing and requirements.

The following diagram depicts the DDR memory controller system's connectivity within the Zynq-7000 SoC:

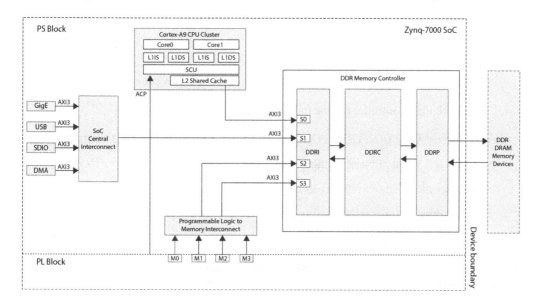

Figure 5.9 – DDR memory controller-centric view of the Zynq-7000 SoC

DDR memory controller features and configuration parameters

The DDR memory controller has the following SoC interface characteristics:

- The AXI interface has the following features:

  - Four ports, each with a 64-bit full duplex data interface with 32-bit addressing capabilities

  - All ports support both AXI INCR and WRAP bursts

  - A multi-port arbitration protocol that maximizes DRAM bandwidth utilization

  - Arbitration logic that can be bypassed for low-latency transactions

  - High acceptance capability per port for both read and write commands

  - Out-of-order read completion for read transactions with different master IDs and an AXI ID width of 9 for all ports

  - 1 to 16 beats burst length, with a burst size of 1, 2, 4, or 8 bytes per beat

  - Programmable secure regions on 64 MB boundaries using a system-level register

  - Exclusive accesses are only supported for two different master IDs and only per port, not across ports

- The DDRC controller has the following features:

  - DDR DRAM memory data throughput and low latency-aware transaction scheduling algorithm (a three-stage arbitration and command selection mechanism, as shown in *Figure 5.12*)

  - A transaction reordering engine to optimize the DDR DRAM access bandwidth for both sequential and random access

  - Read and write address dependency detection for data access coherency

- The DDRP PHY controller has the following features:

  - Supported DDR I/Os: 1.2V LPDDR2, 1.8V DDR2, 1.5V DDR3, and 1.35V DDR3L

  - 16-bit and 32-bit physical memory data width supported

  - Software controller self-refresh mode with automatic exit when there are commands to execute

  - Programmable idle periods timing for the automatic DDR power down entry and exit

  - Automatic calibration for the data read strobe

Port S0 of the DDR memory controller is dedicated to the Cortex-A9 L2 cache interface and the ACP. On the DDRI side, this port is usually configured for low latency access.

Port S1 serves the other PS masters via the SoC central interconnect.

Ports S2 and S3 serve the programmable logic side of the SoC via a switching interconnect. On the DDR memory controller side, these ports are usually configured for high throughput.

The DDR memory controller has 1 GB of address space allocated to it in the 4 GB SoC address map. The following table shows the possible physical devices supported and their arrangements:

Device	Component Configuration	Number of Components	Component Density (GB)	Total Width	Total Density
DDR3/DDR3L	x16	2	4	32	1 GB
DDR2	x8	4	2	32	1 GB
LPDDR2	x32	1	2	32	256 MB
LPDDR2	x16	2	4	32	1 GB
LPDDR2	x16	1	2	16	256 MB

Table 5.2 – Zynq-7000 SoC DDR memory controller example DRAM memory organization

Quality of Service (QoS) is implemented in the DDR memory controller's DDRC using a combination of the round-robin arbitration algorithm among the four port queues and an aging mechanism assigned to each queue via a software-controlled register. There is also an input signal representing an *urgent request* on a port that resets the aging counter, making it the highest priority queue. The arbitration algorithm also uses page matching to decide which command queue to service. The page, which is the DDR memory page, is open at the time of the arbitration. The following diagram illustrates the arbitration algorithm (stage-1) that's used by the DDRC to arbitrate between the ports:

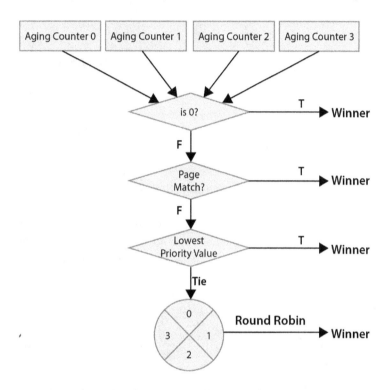

Figure 5.10 – DDR memory controller ports arbitration algorithm

Within the DDRC core, there is another stage of arbitration (stage-2) between the read and the write command queues. The following diagram illustrates this stage of arbitration in the Zynq-7000 SoC DDR memory controller:

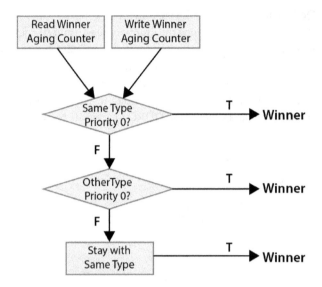

Figure 5.11 – DDR memory controller read and write queues arbitration algorithm

In this second stage of arbitration within the DDRC, the aging counter is checked. Then, an older one is selected. If the next command from it has a priority of 0 and is like the previous one (already selected in the previous arbitration cycle), then it wins the arbitration. For example, assuming that in the previous cycle, the command was a read transaction, the queue that is the oldest is the read queue, and the read command has a priority of 0, then it is the winner of this arbitration cycle, and the next read transaction is elected to be executed next. If it is not the same type of command with a priority of 0, and another type of transaction with a priority of 0 is next, then that one will win. If there is no transaction with a priority of 0 in the queues, then the winning transaction is from the same queue as the previous transaction.

It is important to understand these arbitration schemes to figure out which stream of commands will be serviced next and how such a scheme may affect the overall performance of the system, especially in applications where the overall software processing architecture relies on a pipelined model. This will help in choosing which QoS attributes to assign to a port that's used for a specific data stream.

The DDRC also has an optional feature that is called **high priority read** (**HPR**). Here, the read data queue is split into two separate queues – one for high-priority transactions and another for low-priority transactions. Every port of the DDR memory controller can be qualified as a high or low priority according to the application needs and the nature of the data streams it services – for example, assigning a high-priority read to port 0, which is servicing the CPU. This feature needs to be enabled to be used. The following diagram provides an overview of all the arbitration stages a command goes through before being executed by the DDRC and forwarded to the DDR DRAM memory device:

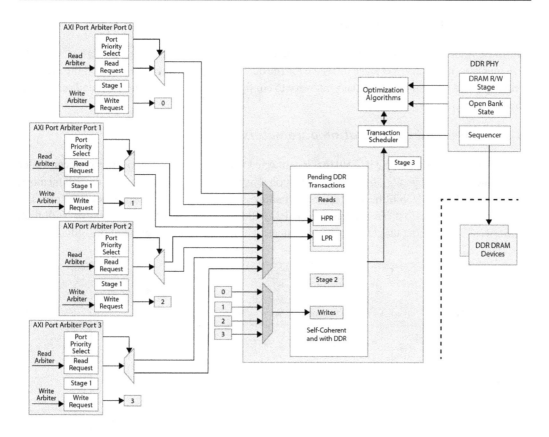

Figure 5.12 – DDR memory controller commands full arbitration stages

The DDRC also has a write combining feature, which can be enabled to exploit merging many write commands to the same address from different commands.

There are many other operational details related to the DDR memory controller, such as the clocking, the reset, the startup sequence, and the controller calibration, which we will cover in *Part 2* of this book when we put an SoC project together that uses the DDR SDRAM as its system's main memory. We will also practically assess the effect certain parameters have on the system's overall performance by putting together a benchmarking software example that can measure the effects on system performance for the given parameter settings. For details on these, you are invited to examine *Chapter 10* of the *Zynq-7000 SoC Technical Reference Manual* at https://docs.xilinx.com/v/u/en-US/ug585-Zynq-7000-TRM.

Static memory controller

The **static memory controller** (**SMC**) can be used to interface to a NAND flash, SRAM, or NOR flash memory. It has an AXI3 slave interface for the data transfers between the SoC and the memory media attached to it, and an APB slave interface to configure the registers of the SMC and read status when necessary.

SMC features and configuration parameters

The SMC has an interrupt (IRQ ID# 50) that is connected to the Cortex-A9 **generic interrupt controller** (**GIC**). The SMC can operate either as a NAND flash interface and address up to 1 GB of flash memory, or as a parallel SRAM/NOR flash interface and address up to 32 MB per **chip select** (**CS**) pin. The SMC has two CS pins, so it can address two dual 32 MB regions of attached SRAM and/or NOR flash memories. The following diagram provides a microarchitectural representation of the SMC:

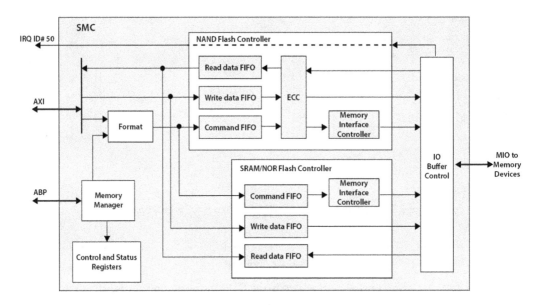

Figure 5.13 – SMC microarchitecture diagram

The SMC NOR or NAND flash memory can be used as a booting device for the Cortex-A9 CPU. For details on clocking and reset options, the external connectivity, the APB registers file, and other operational information, you are invited to explore *Chapter 11* of *The Zynq-7000 SoC Technical Reference Manual* at https://docs.xilinx.com/v/u/en-US/ug585-Zynq-7000-TRM.

On-chip memory controller

The OCM controller is an interface within the PS that has an internal SRAM of 256 KB and a BootROM of 128 KB. The OCM controller presents two AXI3 64-bit interfaces to the SoC masters, one connected to the Cortex-A9 CPU and the other shared between all the other SoC masters (from the PS and PL) via a memory interconnect. The following diagram shows the system view of the OCM controller in the PS block:

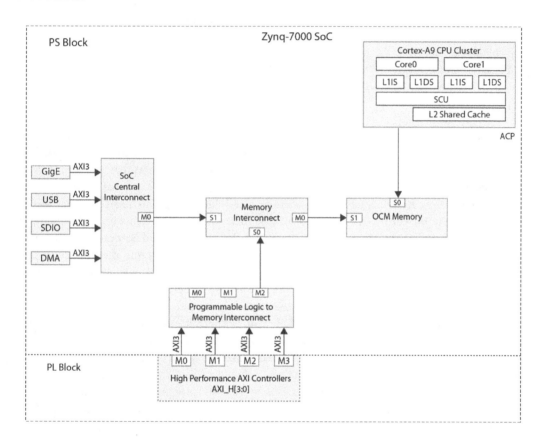

Figure 5.14 – On-chip memory controller-centric view of the Zynq-7000 SoC

OMC features and configuration parameters

The OCM SRAM is divided into 64 x 4 KB pages for which the security attributes (access rights and modes) can be assigned on a per-page basis. The OCM also has a configuration APB slave port and an interrupt signal with ID#35. The CPU/ACP interface to the OCM is on a low latency path but still has a minimum of 23 cycles for read latency when the CPU is running at 667 MHz. Arbitration between the AXI read and write channels is done round-robin from the OCM interconnect, and the arbitration between the CPU/ACP port and the OCM interconnect ports is fixed with the highest priority to the CPU/ACP one. The OCM, due to its lowest access latency from the CPU perspective compared to the DDR DRAM, can be used to host the critical parts of the SoC software, such as the TLB tables, the interrupt service routines, the exceptions code, and any functions that are real-time in nature and greatly affect the system's response time. The OCM is also good for hosting the DMA engine descriptors, as well as any other parts of the code or data that are bound by real-time constraints for it to be accessed. Further details on OCM clocking, its address mapping, and the address relocation options will be covered in *Part 2* of this book when we configure the OCM for critical code and data section hosting. Information on the APB registers file, as well as other operational information, is available in *Chapter 29* of the *Zynq-7000 SoC Technical Reference Manual* at `https://docs.xilinx.com/v/u/en-US/ug585-Zynq-7000-TRM`.

Summary

This chapter started by providing an architectural and functional overview of what SoC interfaces are while classifying them by functionality. Then, we listed the main services needed to perform such an interfacing function. After that, we provided a summary of the main features that globally affect the SoC's system performance that need to be considered when designing an SoC interface of a specific type. Then, we looked at the fundamentals of the processor caches while focusing on the ARMv7 architecture implementation in the Cortex-A9 CPU and covered their associated terminology and topologies. After that, we looked at the cache organizations, the possible mappings, and how caches improve the SoC system performance. We also introduced the processor MMU and its role in virtual to physical address translation, physical address space management, and SoC security implementation. Next, we provided a detailed introduction to the DDR memory controller as being one of the fundamental SoC interfaces where the software application and the system's main data are stored. Here, we looked at the multi-port nature of the Zynq-7000 SoC DDR memory controller and how transactions from the different ports are arbitrated for access to the memory media at the different stages of their processing by the DDR memory controller. We also introduced the static memory controller and the different types of memory and flash devices that can be connected to the SoC through it, as well as its main architectural features and performance metrics. Finally, we introduced the OCM controller as a major complementary storage interface for the DDR memory controller for hosting the critical parts of the SoC software and as the default BootROM of the PS.

This was the closing chapter of *Part 1* of this book, where we mainly focused on the fundamentals and the basic knowledge you need to start designing and implementing an SoC using Xilinx FPGAs. The next chapter will be the opening chapter of *Part 2* of this book, where we will look at the SoC architecture design and development process before moving on to putting the SoC together and building, verifying, and integrating all its critical components.

Questions

Answer the following questions to test your knowledge of this chapter:

1. What is an SoC interface? How can we classify one?

2. How does classifying the SoC interfaces help with designing a custom SoC interface?

3. Build a table where you list the various interface types we can distinguish, and provide the main services for each.

4. How could we compute the depth of a write/transmit FIFO for an SoC interface if we are given the read IC of the AXI3 port of the interface as 16, the speed of the SoC interconnect as 400 MHz, the type of the interface as 5, and the external domain as having a transmission speed of 250 Mbps?

5. How is a processor cache organized? Define all the parameters that are used in the Cortex-A9 cache topology.

6. What is a cache tag? How is its length computed?

7. What is a set-associative cache? What are the main advantages of using a set-associative cache versus using a direct mapped cache?

8. How many ways do we usually find in an L1 ARMv7 implemented cache? What about L2 caches?

9. What is a processor MMU?

10. What is a processor virtual address?

11. How does a virtual address get translated into a physical address?

12. How many ports does the DDR memory controller of the Zynq-7000 SoC have? Why is this topology used?

13. How many transaction command arbitration stages are there in the DDR memory controller of the Zynq-7000 SoC? Explain the principle of each and how the possible choices can affect the system's performance.

14. What are the main features of the Zynq-7000 SoC static memory controller?

15. What types of memory are used in the OCM controller? Explain the system uses of each.

16. Why is the OCM controller's CPU/ACP port given the highest arbitration priority for its AXI3 read sides among all the other ports?

Part 2: Implementing High-Speed SoC Designs in an FPGA

This part introduces the design flow and the tools used, from the architectural concept to the implementation of an SoC design in an FPGA for both the hardware and the accompanying software.

This part comprises the following chapters:

6

What Goes Where in a High-Speed SoC Design

In this opening chapter of *Part 2* of this book, you will learn about the SoC architecture definition phase, which precedes the design and implementation phase of the required SoC. This phase is performed by the system architects, who translate a certain set of product requirements into a high-level description of the SoC design to accomplish. We will also detail the criteria that will be used during the functional decomposition stage in which a trade-off is reached between what is better suited for implementation in hardware and what is rather a good target for software implementation. Finally, we will provide an overview of the SoC system modeling, which can use many available tools and environments.

In this chapter, we're going to cover the following main topics:

- The SoC architecture exploration phase
- SoC hardware and software partitioning
- Hardware and software interfacing and communication
- Introducing the Semi-Soft algorithm
- Early SoC architecture modeling and the golden model

The SoC architecture exploration phase

This is the beginning of the pure technical stage in a project aiming to design an SoC. Usually, the technology to use isn't specified at this stage, but there could be clear business reasons, as covered in *Chapter 1*, *Introducing FPGA Devices and SoCs*, that put the FPGA as the primary target technology for the SoC to design. These reasons can include (but are not limited to) the following:

- The expected production volume is low.
- The time to market and the product opportunity window are narrow.

- The **non-recurring engineering** (**NRE**) cost of an FPGA technology is within the project's budget.

- In this project, using an ASIC has no competitive advantage. It only provides disadvantages and added project uncertainty and risks.

There could be many other reasons for making the FPGA the best target for the SoC to design, which will then benefit the time to market and flexibilities such a choice offers. At this stage of the thinking process, the architecture to use is also centered around the Zynq-7000 SoC or the Zynq UltraScale+ SoC with a starting **processer subsystem** (**PS**) block. We still need to perform our feasibility study and confirm that the project objectives and the FPGA SoC capabilities are in line. Each FPGA SoC device family has a known set of features, capabilities, and associated device costs. An initial idea about the SoC capabilities to include and the major **intellectual properties** (**IPs**) to design is put together based on the input from the marketing team. The marketing team conducts many interviews with the key target customers of the product under design and gathers the key product requirements. It is at this moment that broad guidelines from the business team are needed to define the cost of the overall solution and product that the company is making. However, the business decision could be delayed slightly until the SoC cost is defined, the performance requirements are refined, and the time to design it is estimated. It is at this stage that the overall integration of the product cost could be revisited to figure out alternative business strategies if this requires further adaptations.

The overall product system architecture definition and the business strategy are outside the scope of this book, so we assume these have already been set up by the time the SoC architecture definition is considered. The remaining tasks are to decide which SoC can provide the required interfaces and deliver the performance of the product. This SoC architecture exploration phase will look at all the possible alternatives and the associated design costs in terms of the SoC hardware and software. Then, it will provide a report to the business decision-makers to approve the optimal choice for the project that meets the company strategy. There could be other ways and methods by which this decision is made according to the company's culture, but this decision process is irrelevant to this book.

Additionally, there is also a need to define how much custom work will be needed to complement the SoC PS block's functionalities, features, and performance. A detailed report on this will be provided as an architecture specification chapter in the following section, where we will cover the SoC hardware and software partitioning stage. It's expected that IPs will be designed for this project, implemented in the **programmable logic** (**PL**) side of the FPGA, and integrated with the PS subsystem of the SoC. This also adds another criterion of choice between the SoCs to use from within the same family or among the available families.

As introduced in *Chapter 1, Introducing FPGA Devices and SoCs*, some SoCs specifically target some applications and industries in terms of their capabilities and features, as well as the availability of certain packages and certifications in their portfolio. This choice should also make sure that enough **input/ output** (**I/O**) is available in the SoC package to use, and that the SoC can physically interface with the neighboring **integrated circuits** (**ICs**) in terms of supported PHYs and inter-device communication protocols.

These interfaces include the ones covered in *Chapter 4, Connecting High-Speed Devices Using Busses and Interconnects*, and *Chapter 5, Basic and Advanced SoC Interfaces*, as well as those briefly introduced in *Chapter 1* of this book, such as the following:

- The **Serial Peripheral Interface (SPI)**
- The **Inter-Integrated Circuit (I2C)**
- The **Peripheral Component Interconnect Express (PCIe)**
- The **Universal Asynchronous Receiver Transmitter (UART)**
- The **General-Purpose Input/Output (GPIO)**
- The Ethernet controller
- The DDR memory controller
- The flash controller

The list should also state which version of the protocol standard these interfaces support, as well as what kind of backward compatibility is available if the revision of the standards and the generations aren't the same between the SoC available interfaces and the neighboring ICs within the electronics board of the product. The physical characteristics of the FPGA device in terms of temperature range, package size, mechanical properties, and all required certifications per the industry vertical targeted by the product need to be considered before we start any architecture design work. For these, the list of device requirements should have been established by the project team at the technology target selection phase, before the SoC architecture design phase. This selection process is also outside the scope of this book. To give you an easier and clearer way to decide between the different Xilinx SoCs covered by this book, four tables (*Tables 6.1 to 6.4*) have been provided in this chapter that summarize their available features.

There are clear differences in the processing capabilities of the Zynq-7000 SoC and the Zynq UltraScale+ SoC PS blocks. There is no Cortex-R processor cluster in the Zynq-7000 SoC FPGAs, but applications that require some form of real-time profile and a deterministic processor type may be built in the PL side of the FPGA using the MicroBlaze processor, as introduced in *Chapter 1*. Then, some RTL integration is needed to interface it to the Cortex-A9 cluster over the ACP or simply over the AXI ports available for bridging from the PS block to the PL block in both directions. We will cover this design methodology and techniques later in this book when we cover the available co-processing methods in both the Zynq SoC families of FPGAs.

SoCs PS processors block features

This subsection lists the PS block processors and their features per FPGA SoC type.

SoC PS Cortex-A CPU

The following table lists the Cortex-A CPU features for both the Zynq-7000 and Zynq UltraScale+ SoC FPGAs:

Feature	Zynq-7000 SoC	Zynq UltraScale+ SoC
Cortex-A cluster type	A9 ARMv7-A	A53 ARMv8-A
Cores per cluster	Single core or cluster of 2 cores	A cluster of 2 or 4 cores
ISA	AArch32, 16-bit, and 32-bit thumb instructions	AArch64, AArch32, and 32-bit thumb instructions
Core performance	2.5 DMIPS/MHz	2.3 DMIPS/MHz
Operation modes	Both SMP and AMP	Both SMP and AMP
L1 caches	L1 instruction cache of 32 KB L1 data cache of 32 KB	L1 instruction cache of 32 KB L1 data cache of 32 KB
L1 instruction cache associativity	4-way set-associative	2-way set-associative
L1 data cache associativity	4-way set-associative	4-way set-associative
L2 cache	L2 shared cache of 512 KB	L2 shared cache of 1024 KB
L2 cache associativity	8-way set-associative	16-way set-associative
SIMD and FPU	NEON	NEON
Accelerator coherency port	ACP	ACP
Core frequency per speed, grade, or device type	(-1): Up to 667 MHz (-2): Up to 766 MHz or 800 MHz (-3): Up to 866 MHz or 1 GHz	(CG): Up to 1.3 GHz (EG): Up to 1.5 GHz (EV): Up to 1.5 GHz
Security	TrustZone	TrustZone and PS SMMU
Interrupts	GIC v1	GIC v2
Debug and trace	CoreSight	CoreSight

Table 6.1 – The SoC PS block's processor features

Memory and storage interfaces

The following table lists the memory controllers and the storage interfaces available in both the Zynq-7000 and Zynq UltraScale+ SoC FPGAs:

Feature	Zynq-7000 SoC	Zynq UltraScale+ SoC
DRAM controller ports	4	6
DRAM controller standards	DDR2/LPDDR2 DDR3/DDR3L	DDR3/DDR3L/LPDDR3 DDR4/LPDDR4
DRAM controller maximum capacity	1 GB	8 GB Up to 16 GB for DDR4
DRAM controller ranks	1	1 and 2
QSPI flash controller	I/O and linear modes	DMA, linear, and SPI modes
OCM controller	256 KB SRAM and 128 KB BootROM	256 KB SRAM
SRAM controller	1	1
NOR flash controller	1	N/A
NAND flash controller	1	1
SD/SDIO controller	1	1
SATA controller	N/A	1

Table 6.2 – The SoC PS block's memory and storage controllers

Communication interfaces

The following table lists the communication interfaces available in both the Zynq-7000 and Zynq UltraScale+ SoC FPGAs:

Interface	Zynq-7000 SoC	Zynq UltraScale+ SoC
USB	Device, host, and OTG	Device, host, and OTG
Ethernet	2x (10/100/1,000 Mbps)	4x (10/100/1,000 Mbps)
SPI	2	2
CAN	2	2
I2C	2	2
PCIe	NA in the PS	Gen2 x1/x2/x4 within the PS
UART	2	2

Table 6.3 – The SoC PS block's communication interfaces

PS block dedicated hardware functions

There are also many dedicated hardware functions built into the PS of both Zynq SoCs, as summarized in the following table:

Feature	Zynq-7000 SoC	Zynq UltraScale+ SoC
DMA	8x channels	8x channels LPD
		8x channels FPD
ADC	1x XADC	1x SYSMON
GPU	N/A	ARM Mali-400 MP2
PMU	N/A	MicroBlaze-based PMU
Display controller	N/A	1x VESA DisplayPort v1.2a

Table 6.4 – The SoC PS block's dedicated hardware functions

FPGA SoC device general characteristics

In the SoC architecture exploration phase, we also need to know the following:

- What are the available densities in terms of PL logic elements?
- Which packages are available for the target FPGA SoC?
- What is the maximum I/O offered per package?
- What are the temperature ranges for a specific FPGA SoC package?
- What are the speed grades of the SoC FPGAs that we can potentially target?

The speed grade classifies the FPGA SoCs in terms of the maximum frequency the design elements' PS and PL can run at; there is a higher price tag attached to a higher FPGA SoC speed grade. These details are too vast to summarize here, but all these are provided by Xilinx in the FPGAs selection guides documentation. You are encouraged to read the *Zynq-7000 SoC FPGAs Product Selection Guide* at https://docs.xilinx.com/v/u/en-US/zynq-7000-product-selection-guide and the *Zynq UltraScale+ SoC FPGAs Product Selection Guide* at https://docs.xilinx.com/v/u/en-US/zynq-ultrascale-plus-product-selection-guide to learn more.

The key defining elements of the specific FPGA SoC to choose are the PS block's required processing power and the amount of FPGA logic elements necessary to build the custom hardware that implements the key acceleration functions or the company IPs. It is logical to start with the most cost-effective option and build on it by adding more features as more of the details are unveiled, thus moving on to the next target device. Since the hardware and software partitioning phase hasn't been accomplished yet, it is a good idea to perform a technical assessment to estimate the main functions that will be executed in hardware that are impossible to run in software, or that are company or third-party IPs forming part of the overall SoC architecture. Also, any new company IP that is to be built into the hardware will be listed, which will give us an idea in which direction the choice should be heading.

For this chapter, a practical example is the best approach to apply the ideas and suggestions listed thus far since we will be implementing this first simple but complete design in the next chapter. Therefore, we will start with the SoC architecture, which is based on a Zynq-7000 SoC since we would like this exercise to be simple and illustrative.

We need to perform the following tasks:

1. Decide upon the SoC architecture.
2. Perform the hardware and software partitioning.
3. Define the **hardware-to-software** and **software-to-hardware** interfaces and communication protocols.
4. Configure the SoC PS with the necessary IPs for our processor system architecture.
5. Build the custom IP section (if any).
6. Integrate the custom IP into the overall SoC hardware design.
7. Implement the design.
8. Simulate any part of the design that is subject to RTL design flow.
9. Build some test software to verify that the hardware is functioning as expected.
10. Implement the design.
11. Finally, if we have a demo board available, we can use that to check that the design is fully functional. To do so, we can download the configuration binary and executable files.

In *Part 3* of this book, we will be building more complex SoCs that require higher processing power and therefore potentially targeting the Zynq UltraScale+ SoC. We will also look at performing system profiling to help us implement custom acceleration hardware that we will also integrate into the SoC. We will also build the necessary software drivers for these custom functions that will run under an RTOS such as Embedded Linux.

To conclude the architecture exploration phase, we must come up with a list of possibilities that have advantages and disadvantages. By doing this, we can compare them in terms of cost, design, verification effort, and time. We usually discuss these with the business stakeholders and decide on the best option. After this, we start mapping the functions of the SoC on the **processing elements** (**PEs**) that are in either hardware blocks or software functions and start the next phase of the SoC architecture development.

SoC hardware and software partitioning

As mentioned in the previous section, the architecture devolvement task of mapping the functions of the SoC to the PEs available is better exercised using a practical example.

A simple SoC example – an electronic trading system

To perform hardware and software partitioning, let's design an SoC that implements the intelligent parts of a dummy financial **Electronic Trading System** (**ETS**). It's a dummy since it isn't a system that we can use to perform financial transactions in an **Electronic Trading Market** (**ETM**) managed by a specific private organization; it just behaves like one. Most financial ETSs are co-located in a data center managed by a private organization. The interface between the ETM and the trading clients is a network switch where the trading clients plug in their network interfaces, which connect them to the ETM. The market itself is a network of servers that broadcasts the market data over, for example, the **user datagram protocol** (**UDP**) and receives trade transactions and their confirmation over, for example, the **transmission control protocol/internet protocol** (**TCP/IP**). Both UDP and TCP/IP are part of the **Internet Protocol** (**IP**) suite, commonly known as the TCP/IP stack, and are widely used in computer networking and communications architectures. Further details on the TCP/IP protocol suite can be found at `https://datatracker.ietf.org/doc/html/rfc1180`.

In the envisaged ETM wider network architecture, every client is connected to the trading market switch over Ethernet interfaces, and listens to the market information over the UDP broadcasted packets. The information is specific to the ETM itself and is formatted as we see fit. Our system should be able to cope with any formatting used and be able to adapt to it if it's updated by the ETM. The ETM organization uses trading symbols that represent a financial market product. Information about these symbols, such as the asking prices, the volumes, the transactions on a given symbol, and many other details, are broadcasted by the ETM. This information is what the ETM has pre-formatted and what the clients listen to while electronic trading is open. The clients receive the ETM information over UDP via their Ethernet interfaces from the ETM switch, decode its content, filter it, make decisions in software (or accelerated software in hardware), and then inform the market of any buying or selling decisions over their TCP/IP connection. As we can imagine, the client with the fastest round-trip communication and processing is the one who can maximize their gains and be the first to exploit a good opportunity, such as a low asking price for a symbol they are targeting and wish to invest in. This race to zero latency is at the heart of the low-latency and high-frequency ETMs that exploit the superior technological solutions a trading organization may possess to drive the market in one direction or another. The following diagram depicts the simplified electronic financial market concept:

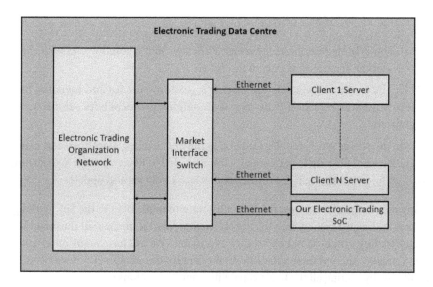

Figure 6.1 – Electronic trading data center concept

For our SoC architecture development exercise, we need to design an SoC that can do the following:

- **[CP1]**: Listen to the market data over the UDP stream using the Ethernet port.
- **[CP2]**: Extract the information from the received Ethernet packets.
- **[CP3]**: Make sure that the UDP packet is valid and that its fields are as expected.
- **[CP4]**: Understand the ETM protocol and act accordingly.
- **[CP5]**: Distinguish between the trading information data and the ETM systems management data that's broadcast over a different UDP stream but over the same Ethernet link.
- **[CP6]**: Extract the information of interest to the software (running on the Cortex-A9) and feed it to our trading algorithm, which is also running in the software.
- **[CP7]**: Consume the data fed via the receiving mechanism and use it to feed our trading algorithm implemented in the software.
- **[CP8]**: Maintain a database where traded symbols of interest are stored alongside the associated information (date, time, volume, and price). This database should be stored in nonvolatile memory.
- **[CP9]**: Make trade decisions when certain conditions are met for a specific trading symbol.
- **[CP10]**: Send the trading decision over the TCP/IP connection to the ETM organization using the Ethernet link to the market interface trading switch.
- **[CP11]**: Maintain a database of its transactions, including the date, time, volume, and price.
- **[CP12]**: The transactions database regarding its trades is also stored in non-volatile memory, but this is encrypted using a private key stored in a secure location.

The electronic trading SoC is part of a trading server:

- **[CP13]**: This is hosted in one of the server PCIe slots and communicates with the server over a PCIe endpoint integrated as part of the SoC.

- **[CP14]**: The trading server sets the trading algorithm policies and manages the databases previously mentioned using PCIe over a predefined protocol between the server and the SoC software.

- **[CP15]**: The server sends regular updates to the SoC trading algorithm and can modify the way UDP packet filtering is performed. This is needed to make sure the SoC trading engine is always up to date with the ETM organization's latest policies and updates.

The preceding list of capabilities is just the bare minimum to implement in the SoC to design the ETS. The objective is to design a system with the lowest latency possible and use all the possible techniques to make such a trading system as fast and secure as possible. We will have many other questions as we design the SoC architecture, but these should be on the details side rather than a fundamental architecture issue. We will start by putting all the listed capabilities and the options we can use to implement them in a table. Then, we will cover the overall implementation for every capability before seeing whether they are better suited for software, hardware, or a combination of both in our low-latency ETS. The following table classifies these capabilities and their possible implementation options:

Capability	Hardware	Software	Both
[CP1]	√	√	√
[CP2]	√	√	√
[CP3]	√	√	√
[CP4]	√	√	√
[CP5]	√	√	√
[CP6]	√	√	√
[CP7]	√	√	√
[CP8]		√	
[CP9]	√	√	√
[CP10]	√	√	√
[CP11]		√	
[CP12]		√	
[CP13]	√	√	√
[CP13]	√	√	√
[CP14]	√	√	√
[CP15]	√	√	√

Table 6.4 – Electronic trading SoC capabilities classification

As you can see, all the capabilities except database management can be implemented in hardware only, software only, or using both software and hardware. We have excluded the database management as it is a background task not worth designing a mechanism for from scratch in RTL. At this stage, we are also looking to assess the effort required to design a capability in hardware since we are assuming that it will be faster when executed in a hardware PE specifically designed for it. Most of the time, this is true. But the real question to ask is whether this speed-up is worth the effort, the time spent, and the implementation cost. To do so, we need to draw a back-to-back data communication pipeline for the ETS, highlight the critical paths in this communication pipeline that are sensitive to time, and understand what it would mean if we were to move a capability from software to execute it in its custom form to be designed by a PE. We are assuming that the easiest implementation (but not necessarily optimal in terms of speed) is putting everything in the software. Once a capability is moved from software to hardware, we can understand what the interface between the two looks like. This is important and by itself requires further consideration.

If we look at our ETS, the back-to-back communication pipeline that is sensitive to latency is from the time a UDP packet hits the Ethernet port of the SoC to the moment a TCP/IP packet with a trade decision is sent from the SoC back to the switch, acting as the interface with the ETM via the same or another Ethernet port. The fastest solution is to design everything in hardware and use IPs that are designed for very low latency, but this is the spirit of a high-frequency trading system and defeats the purpose of our book's objective, which is learning how to design SoCs and integrate PL-based IPs with them using both the PS and PL blocks of the FPGA. We would still like to design a low-latency solution, but we don't want to design an exotic TCP/IP stack and middleware, or any hardware-based **real-time operating system (RTOS)** in RTL. Therefore, we will use a simpler approach where we can use hardware acceleration when it makes sense to meet our design objectives. Our back-to-back data communication pipeline does the following:

- Receives the UDP packet over the Ethernet interface.
- Checks the content of the UDP packet and makes sure it is a valid packet that matches either the management or the market data format, as well as that it has no errors in it by computing the electronic trading protocol error checking code such as the **Cyclic Redundancy Check 32-bit (CRC32)** that's computed over all the packets. The UDP payload transports the ETP, which defines the format of the payload, the length of the packet, a packet number, a timestamp, all the fields along with their length and meaning, and a CRC32 that's computed over the packet and inserted at the end of it. We will revisit the design of the ETP later in this book when we start building, simulating, and integrating the ETS.
- If the content is a *management packet*, then it's sent to the *management queue*, which is associated with a task called **management task** that runs in the software on the Cortex-A9, to deal with it by also notifying it using an interrupt.
- If the content is *market data*, then it filters it using the information set by the trading algorithm, called the **trading algorithm task**, which is running as a high-priority software task on the Cortex-A9.

- If the preceding market data is a symbol our trading algorithm is interested in and for which the filter returned true (such a specific price is lower than the price set in the filter), then the filtering PE should put it in the *urgent buying queue* and notify the *trading algorithm task* via a high-priority interrupt.

- If the preceding market data is a symbol our trading algorithm is interested in and for which the filter returned true (such a specific price is higher than the price set in the filter), then the filtering PE should put it in the *urgent selling queue* and notify the *trading algorithm task* via an even higher priority interrupt.

- All market data symbols of interest to the trading algorithm that met the filter conditions (a symbol matching a set of symbols) are also put in a queue, called the **market data queue**, and sent to the *market database manager task* running in the software on the Cortex-A9 to store them in non-volatile memory. There is also a low-priority interrupt associated with this notification that's sent to the market database manager task.

- If a trade is performed on a target symbol by the trading algorithm task, then the trading algorithm task puts it in a *trade queue* and notifies the *trade database manager task* via a software interrupt to add it to the trade database, but first, it will encrypt it.

- The trade database manager puts the trade information it received from the trading algorithm task securely in an encryption queue that another task will manage as a low-priority task in software, or even as a task subject to offload to the hardware. For this, we need to implement a PE in the FPGA PL block. This is done not to accelerate it, but to free up heavy CPU usage in such a task and leave the performance for urgent tasks, such as dealing with the trading algorithm and interaction with the trading queues. This is subject to profiling to decide what to do and can always be changed at a subsequent design stage as an improvement or an optimization of the CPU usage and its shared resources.

The server side will not be included in our simple example design of this electronic trading SoC, but it may be a good addition to cover PCIe. This will be covered in the advanced applications of this book in *Part 3*.

The following diagram summarizes our electronic trading receive communication path:

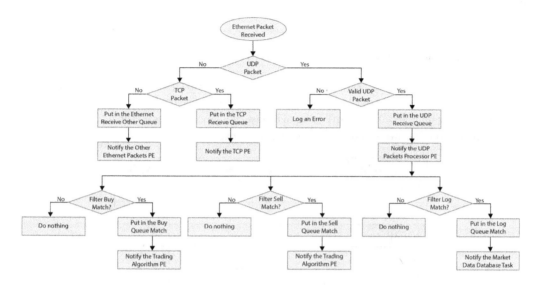

Figure 6.2 – Electronic trading receive communication path

The preceding diagram illustrates the frontend of the electronic trading communication path, where the data of interest for trading can be easily highlighted, analyzed, and then mapped to either a hardware PE, a software PE, or a combination of both. The backend communication path is assumed to be well-adapted to software, so it will be implemented as such. This is why it isn't shown in the preceding diagram. It needs a TCP/IP stack to communicate trades to the ETM. The TCP/IP protocol is an enormous task to implement in hardware, so it's not worth the cost and effort to find out whether it is available as a third-party IP to license for our project. The latency gains it may give us aren't important for this specific ETS. The software trading algorithm has three high-priority tasks that run on the Cortex-A9:

- The *Sell task*, which consumes entries from the *urgent selling queue*.

- The *Buy task*, which consumes entries from the *urgent buying queue*.

- The third task is to *interface with the server*, via which algorithm policy changes are forwarded to the electronic trading algorithm from the traders.

As mentioned previously, we are not focusing on the server side of the system in this architecture definition example.

Now, let's analyze these paths and highlight the critical paths in this ETS to evaluate which PE type or combination will be used to implement them.

From *Figure 6.2* and the previous descriptions, we can conclude that the critical paths for our low-latency trading SoC are as follows:

- **[Path 1]**: Start at the receiving end of the UDP packet to make a sell decision.
- **[Path 2]**: Start at the receiving end of the UDP packet to make a buy decision.

Therefore, to make a low-latency solution, all the corresponding PEs required to fulfill the tasks involved in these two critical paths for our low-latency electronic trading system need to be implemented in hardware; putting them in software isn't going to be fast enough. Now, let's revisit *Table 6.4* and update it so that it suits our hardware and software partitioning exercise:

Capability	Hardware	Software	Both
[CP1]	√		
[CP2]	√		
[CP3]	√		
[CP4]	√		
[CP5]	√		
[CP6]	√		
[CP7]	√		
[CP8]		√	
[CP9]	√		
[CP10]	√	√	√
[CP11]		√	
[CP12]		√	

Table 6.5 – ETS capabilities partitioned between the hardware and software PEs

Please note that we have removed [CP13], [CP14], and [CP15] from the table as we are not designing the Server PCIe integration side of the SoC in this initial architecture design example.

We still need to define how we will be managing the split between hardware PEs and their peer software PEs, and what the interfacing between them should be. We also need to implement a way by which Ethernet packets that aren't of interest to our hardware acceleration paths are returned to be consumed by the software directly, since that would be done if no hardware acceleration was implemented. We also need to study the consequences of our partitioning on the Ethernet controller software driver and make the necessary changes to adapt it to our new hardware design. All these important details will be covered in the following section.

Hardware and software interfacing and communication

The acceleration path introduced in the end-to-end communication path between the ETM switch and the software running on the Cortex-A9 is only necessary for the UDP packets received over the Ethernet port. Anything else that's received over this Ethernet port, such as Ethernet management frames and ARP frames, we have no interest in accelerating, so we would like the acceleration path to be transparent to them. However, we can't do this by simply returning the received Ethernet packets that aren't of interest to our acceleration hardware to the Ethernet controller. This is because the received buffer within the Ethernet controller is a FIFO that the Ethernet frames can't be written back to once they've been consumed from it. One approach would be to let the Cortex-A9 processor perform the Ethernet frames receive management and pass them to the hardware acceleration PE, which behaves like a packet processor. This is the best approach that will introduce less work for the implementation and specifically the Ethernet controller software driver, as well as the mechanism by which the hardware acceleration PE is notified of the arrival of new frames. The hardware DMA interrupt is hard-wired to the Cortex-A9 GIC, and as such the easy way to pass notifications is via the Cortex-A9, which will be acting as a proxy in this respect.

Data path models of the ETS

In the data reception model of the Ethernet controller, which is using its DMA in the Zynq-7000 SoC, the Cortex-A9 software sets the DMA engine within the Ethernet controller to transfer the received Ethernet frames to a destination memory. Then, the DMA engine notifies the Cortex-A9 via an interrupt. We want this Ethernet frame transfer to be done in memory located within the PL or to the OCM or the DDR DRAM memory. When the Cortex-A9 receives an interrupt from the Ethernet DMA engine when the Ethernet frames received are transferred to the nominated memory, the Cortex-A9 rings a doorbell register within the hardware accelerator domain to notify it that several Ethernet frames have been received. It also tells it how many of them have been received. First, we want to filter these Ethernet frames, extract the UDP frames from them, and then put back the other Ethernet frames where the Cortex-A9 is expecting them. It is only after we filter in the hardware acceleration engine and store it in the receive memory of the non-UDP packets that a second notification is sent to the Cortex-A9 via another interrupt. This subsequent interrupt will tell the Cortex-A9 that Ethernet frames have been received that it needs to deal with. The first DMA receive interrupt to the Cortex-A9 was just for the Cortex-A9 to forward them due to the hardware filtering and processing PE. Breaking the received data path model using the Ethernet DMA engine and its associated interrupt notification still has another aspect that we must deal with, which is the data exchange interface mechanism between the Ethernet DMA engine and the Cortex-A9 software. This exchange is done via the DMA descriptors that are prepared by software for the DMA engine to use when data is received by the Ethernet controller. As mentioned in *Chapter 4, Connecting High-Speed Devices Using Busses and Interconnects*, these descriptors specify the data local destination and the next pointer of the descriptor, as well as an important field known as the **Ownership** field. The Ownership field tells the DMA engine that the DMA descriptor is valid and has been consumed by the software from its previous use and that it is ready to be reused again.

We need to make sure that the task that's recycling the DMA descriptors after consuming their associated received data is performing this task properly and performed for the UDP packets that have now been consumed by the hardware PE, not the Cortex-A9 software anymore. This is fine as we simply need to include this mechanism between the hardware PE and the **DMA Descriptors Recycling Task (DDRT)** in software via a **DMA Descriptor Recycling Queue (DDRQ)** via which the hardware accelerator engine sends requests to the DDRT running in software on the Cortex-A9. Another issue we have introduced in this reception model over the DMA engine of the Ethernet frames is that the consumer of the Ethernet frames is not only the Cortex-A9 but both the hardware acceleration engine and the Cortex-A9. This model introduces an out-of-order consumption of the Ethernet frames, but this isn't a problem as the out-of-order would have been noticeable if it introduced some discrepancy in the DMA descriptors reuse model, which should have enough entries to make this reordering fine. That is, by the time we get to reuse a continuous set of DMA descriptors for subsequent receive operations, both the software and hardware would have been filled with their associated data. Even if the hardware had finished first with its DMA descriptors flipping the Ownership field, the software would have had time to catch up.

If there is a dependency between the received Ethernet frames, this shouldn't be an issue as the ETP guarantees that no change is introduced in the protocol until all the clients have acknowledged the reception of its update and have adjusted to it. This condition requires having a large enough DMA descriptors pool that the slowest consumer is allowed to finish while the Ethernet controller keeps up with the speed at which the Ethernet frames are arriving. This can easily be computed using the maximum rate at which we expect the Ethernet frames to be arriving. The Ethernet interface's default receive path includes the controller hardware, the associated receive DMA engine, and the Ethernet software drivers. To minimize the changes in the receive path, which now includes the added UDP packets filtering, the hardware acceleration engine will deal with the filtering task on a set basis. It will be sending a job completion notification to the Cortex-A9 when it has consumed all the Ethernet frames that have been received within the last set. When the hardware acceleration engine finds a UDP packet within the received Ethernet frames set from the ETM, it does the following:

- It processes the Ethernet frame.
- It puts its DMA descriptor in the DDRQ for the DDRT to be marked as consumed (returned to the pool of DMA descriptors).
- It sends the notification to the Cortex-A9 processor when it reaches the last frame in the last received set.

All non-treated Ethernet frames (which are not UDP packets) are left for the Cortex-A9 to consume and are left as follows:

- They are still in situ in memory like when they were put there by the Ethernet DMA receive engine.
- Their DMA descriptors are still marked as owned by the DMA engine, which means that the Cortex-A9 can consume them.

- When processed by the Cortex-A9 processor, it will flip the **Ownership** field of the DMA descriptor to mark it as recyclable.

In this model, we are keeping everything looking almost the same in the flow of processing, without hardware acceleration of the UDP packet processing. Here, we are just delaying the processing of the non-UDP packets by the Cortex-A9 until the hardware acceleration engine has had the chance to inspect the received Ethernet packet, deal with the ones found to be market data UDP packets, notify the market treading tasks about the UDP packets of interest, and request the DDRT to mark the DMA descriptors of these frames as consumed before sending a final notification to the Cortex-A9 so that it can deal with the remaining non-UDP packets, if any, recycle their corresponding DMA descriptors back to the pool, and call the Ethernet driver to finish the receive flow.

The changes in the hardware model will require some associated changes to be made in the Ethernet controller software model, but this can easily be done since the Ethernet receive path can function in delayed mode. Therefore, the changes we are introducing by adding the filtering path can be added almost silently from the Ethernet controller driver's perspective. It is just what the Cortex-A9 does now when receiving the Ethernet DMA receive notification that has changed from calling the Ethernet frame processing to passing them to the hardware PE, and then waiting for the hardware PE notification to start processing any remaining received Ethernet frames that weren't accelerated. It is here that the UDP that was found to be urgent and that matches the filters is dealt with by the hardware PE. To summarize, the hardware-to-software interfacing and communication process goes as follows:

1. The software sets the Ethernet receive DMA engine's DMA descriptors to use a destination memory to be within the PL block, OCM, or DDR memory.

2. The software waits for the Ethernet receive DMA engine's interrupt.

3. The software prepares a request for acceleration by filling in information about the DMA descriptors to use and the number of receive packets.

4. The software notifies the hardware accelerator engine that there are received Ethernet frames within the memory that need to be filtered and processed with UDPs.

5. The hardware accelerator engine processes the packets of the Ethernet frames that are found to be UDP packets from the electronic trading market.

6. The hardware accelerator populates the urgent buying queue and notifies the trading algorithm task via an interrupt when it finds a matching symbol.

7. The hardware accelerator puts the DMA descriptor associated with the UDP packets in the DDRQ for the DDRT to recycle them.

8. The hardware accelerator populates the urgent selling queue and notifies the trading algorithm task via an interrupt when it finds a matching symbol.

9. The hardware accelerator puts the DMA descriptor associated with the UDP packets in the DDRQ for the DDRT to recycle them.

10. The hardware accelerator populates the market data queue and sends a notification to the market database manager task via an interrupt when it finds a matching symbol.

11. The hardware accelerator puts the DMA descriptor associated with the UDP packets in the DDRQ for the DDRT to recycle them.

12. Once the Ethernet frames have all been inspected, the hardware accelerator notifies the Cortex-A9 via an interrupt that it has processed all the Ethernet frames from the last ones it has been asked to deal with.

13. The Cortex-A9 checks the received Ethernet frames to check whether any Ethernet frames haven't been dealt with yet. If there are, it consumes them.

14. The Cortex-A9 flips the Ownership field of the DMA descriptors for the Ethernet frames that it dealt with itself.

15. The Cortex-A9 notifies the Ethernet drivers or the DDRT to perform any DMA descriptor recycling actions.

The following diagram illustrates these steps:

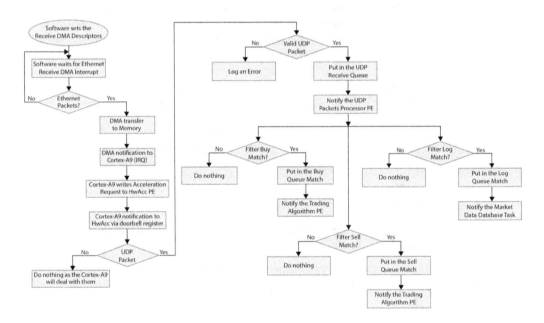

Figure 6.3 – ETS low-latency path hardware to software interaction

Introducing the Semi-Soft algorithm

The Semi-Soft algorithm idea isn't new, and it has been around for many decades now since FPGA technology became prevalent in the electronics industry. It is a combination of hardware and software from the initial architecture development stages and is used to implement compute algorithms.

Using the Semi-Soft algorithm approach in the Zynq-based SoCs

When targeting a Zynq SoC, the hardware and software split between the PEs is considered from the start of the project rather than all the computing algorithms in the software being implemented, profiled, having the bottlenecks pinpointed (if any), and then hardware-accelerated. There is nothing wrong with the latter approach – it is just limiting in what can be achieved when targeting an ASIC technology to implement the design. Once the overall SoC architecture and microarchitecture have been defined, it is hard to modify the design at a later stage and introduce the optimal communication paths between the hardware and the software. When the SoC targets an FPGA, as in our case, the simpler approach should consider the acceleration from the architecture phase, as we have, so that we can put packet processing compute algorithms in place that are Semi-Soft. This means we can modify them easily in both the hardware and the software, so long as we have the appropriate interfacing between the two.

A Semi-Soft algorithm is a computational methodology that uses both the hardware and the software from the start of the architecture design phase. In our example, we are focusing on packet processing, which can greatly benefit from the Ethernet packets being inspected in parallel. Since all the UPD fields can be checked in parallel, many packets can also be checked simultaneously, and many decisions and results can be found and forwarded to the Cortex-A9 in parallel. Although the Cortex-A9 will have to deal with them sequentially when using a single-core CPU, it is capable of processing them in parallel if we deploy a dual-core CPU cluster, where each core is dedicated to a trade (buy or sell) queue.

Consequently, we continue maximizing parallelism and making faster trade decisions. The possibilities are enormous when we consider the compute as a mixture of generic software sequential compute resources of the PS and the parallel and multi-instance possible hardware acceleration engines of the PL. These could augment the PS capabilities to make a powerful custom hybrid processor. In our ETS, the Cortex-A9 manages the Ethernet DMA engine, recycles the DMA descriptors on behalf of both the hardware and software, and runs the middleware for the TCP/IP stack and other Ethernet link management. On the other hand, the hardware performs the Ethernet frame filtering and UDP packet processing, and then notifies the software to complete any remaining packet processing that does not need to be low latency or is complex to perform in hardware.

Using system-level alternative solutions

A design methodology that has a flexible way of moving PE elements in and out of the PL to replace elements from the software tasks or to be implemented instead in software can be matched with other, more complex, hardware electronic systems. One example is when using a PCIe interconnect hosting a hardware accelerator such as a GPU, which can be used to accelerate parallel compute operations for the main processor. We can think of another architecture that uses discrete electronic components that can do the packet processing task. A **network processor** (**NP**) hosted over the PCIe interconnect of a server platform can perform the packet processing tasks, but as can be imagined, first, the cost is at least an order of magnitude higher, the design framework is complex, and the required technical skills to put such a system together is also demanding in comparison to using an SoC-based FPGA. Also, using a server-based PCIe approach won't be low latency; it may be a good approach for high-volume traffic processing, but it's not necessarily comparable for low latency. In this architecture, we have added more layers to the communication stack between the software and hardware accelerators. Here, the platform can only efficiently perform packet processing if we need to accelerate another aspect of the software tasks that were found to be inefficient in the software via system profiling. The FPGA SoC provides us with flexibility and a paradigm shift toward the architecture design phase via an early system implementation that we call the Semi-Soft algorithm concept.

Introduction to OpenCL

Other solutions provide a whole framework for performing this mixing by defining hardware functions and integrating them into a kernel, such as the OpenCL approach. However, OpenCL tries to abstract compute operations from the compute engines performing it and defines methods to match them. You can find out more about OpenCL at `https://www.khronos.org/opencl/`.

Exploring FPGA partial reconfiguration as an alternative method

Another methodology that was also used in the past is to exploit the FPGA partial reconfiguration feature, where a block of the FPGA is defined as a reprogrammable function and used on demand by software running on the same FPGA or externally and interfaced to the FPGA. This means that the design can define many computational functions that can fit within this reprogrammable block of the FPGA, and when a computation acceleration is required, the reprogrammable block is reconfigured to perform this specific hardware acceleration task. This reprogrammable block should have a predefined interface with software through which data and commands are provided to the hardware function by software, and results and statuses are generated by the hardware accelerator hosted in this reprogrammable block.

More information on the FPGA partial reconfiguration can be found at `https://docs.xilinx.com/v/u/2018.1-English/ug909-vivado-partial-reconfiguration`.

Early SoC architecture modeling and the golden model

When targeting an FPGA for an SoC implementation, reimplementing the design takes a matter of days or even sometimes just hours. It involves changing the behavior of an RTL block or drop in a verified IP. Doing so when the target technology is an ASIC is a lengthy process with significant costs in terms of resources and budget.

We will introduce system modeling in the closing section of this chapter as it is part of the architecture development in general and it is also becoming a time-to-market solution for system implementations that take a long time to accomplish, specifically when targeting an ASIC technology. The industry is exploiting the availability of detailed system models of processors, interconnects, and all the IP elements of an SoC to build virtual SoCs. These are system models that emulate the functional behavior of an SoC in software simulation. Many frameworks are available that can put system models together. These system models behave functionally like the target hardware SoC, and software can be prototyped on them earlier than the SoC hardware and electronics board availability. Timing accuracy is sometimes sacrificed to accelerate the timely execution of the system model in the simulation itself when the system model is used for early software development. In this case, we produce a version that is called a **virtual prototype** (**VP**); quantum or simulation time is sometimes fast forwarded to areas of the prototyped software execution of interest to the developer.

The major frameworks of system modeling are as follows:

- System modeling using Accellera SystemC and TLM2.0
- System modeling using Synopsys Platform Architect
- System modeling using the gem5 framework
- System modeling using the QEMU framework and SystemC/TLM2.0

The final timing accurate system model is what we call a golden model. It is usually built according to the SoC architecture specification. This can be checked against the system's final verification so that it can be benchmarked for its correct functionality and achieved performance.

System modeling using Accellera SystemC and TLM2.0

Many IP vendors provide SystemC and TLM2.0 models of their IPs so that they can be used in system modeling to perform the early architecture exploration activities and build VPs for early software development. As mentioned previously, this captures the system architecture specification and helps make a golden model against which the functional verification of the system's RTL can be compared.

SystemC and TLM2.0 are becoming one of the de facto standards for building IP models and connecting them in a transaction-oriented way using TLM2.0 socket-based interfaces. SystemC and TLM2.0 are available for free from Accellera, which also provides the SystemC basic simulator. For more information on this framework, go to `https://www.accellera.org/downloads/standards/systemc`.

This framework is known as the **Open SystemC Initiative** (**OSCI**), which is now owned and maintained by Accellera. It is an **Institute of Electrical and Electronics Engineers** (**IEEE**) standard under IEEE 1666-2011. SystemC is a C++ class library and provides macros that allow you to build concurrent processes that can communicate between themselves through function calls and returns. It is also associated with **transaction-level modeling** (**TLM**), a socket-based programming methodology also based on the C++ library and macros. It provides many ways of mimicking handshaking data and controlling information exchange protocols and helps in building SoC interconnects. While this framework is available for free and can be used to build initial system models, there is a lack of availability for free specific processors and interconnect IPs system models, so the IP vendor needs to be contacted in case they have them available for licensing.

Accellera (or OSCI) SystemC/TLM2.0 is great for building system IP models if we are using RTL, so it provides a rapid prototyping methodology for custom IPs. It is also very useful in building system-level prototypes where no specific CPU ISA is required as it has a SystemC/TLM2.0 model of a generic RISC CPU. This can be used with Assembly language for sequential operations, but this will only allow accurate system modeling that isn't CPU-centric, which means it's not very useful for SoC VPs on their own. However, it can be combined with other frameworks as most of the VPs have a bridge to SystemC/TLM2.0, which means they can still be part of the tools a system architect uses to build a final system model of the targeted SoC with freely available tools.

System modeling using Synopsys Platform Architect

Platform Architect is a **graphical user interface (GUI)**-based system modeling environment from Synopsys. It is based on the SystemC/TLM2.0 framework but has a lot of utilities for making the system modeling and the architecture exploration tasks easier than it is on the OSCI framework. Synopsys has built many system models for their IPs and integrated many third-party IPs within Platform Architect, which can be dragged and dropped in the GUI to stitch a system model together. This comes with a significant price tag, so some third-party IP models – specifically, the CPUs and interconnects – need an additional license from their providers to be able to use them in any system modeling work. For major SoC projects targeting ASICs, Synopsys Platform Architect could be a good solution, but for FPGA-based SoCs, other alternatives for system modeling should be considered.

Further information on the Platform Architect framework is available at `https://www.synopsys.com/verification/virtual-prototyping/platform-architect.html`.

System modeling using the gem5 framework

gem5 is used in computer architecture research and is a framework for performing multi-processor system simulation, which it does by assembling computer system elements. The framework is centered on the CPU models that are built into C++ and configured using a Python script. This framework is suitable for SoC system modeling as it is also a method by which SoC component models such as processor cores, interconnect, memory interfaces, and peripherals can be connected using the configuration Python script to form a custom SoC emulating the real SoC hardware. Custom IPs can be written to extend the portfolio of gem5 and allow the user to produce a system model for the SoC target architecture. gem5 is free and provided under a BSD-like license. A full system can be built, including hardware, operating systems, and application software, to be run on the gem5 simulator. It supports two operating modes:

- **Full System (FS)** mode
- **Syscall Emulation (SE)** mode

For the SoC architecture development tasks, FS mode is suitable for system modeling, whereas SE mode is better suited for early software prototyping and software-centered research. CPU models aren't an exact RTL translation of their architecture implementation, but it uses an execution model with support of the specific **Instruction Set Architecture (ISA)**, which is a good enough execution environment for architecture exploration-associated work. Using gem5 requires a good understanding of the CPU microarchitecture, including its internal memory hierarchy and its settings to customize the CPU using the Python script. This approximates it to a system model that emulates the targeted CPU architecture to be used within the SoC. SystemC/TLM2.0 models are supported by gem5 either by bridging from gem5 to SystemC/TLM2.0 using a full bridge model available in gem5 or by running the SystemC/TLM models within the built-in SystemC/TLM2.0 simulation kernel. This last method is near native in the integration and configuration of SystemC/TLM2.0 models since they can be integrated and configured into the SoC system model using Python, such as the native IP models of gem5. More details on the gem5 simulator can be found at `https://www.gem5.org/`.

System modeling using the QEMU framework and SystemC/TLM2.0

QEMU is a machine emulator with multiple operating modes:

- User-mode emulation
- System emulation
- KVM hosting
- Xen hosting

System emulation mode is the one that's relevant to SoC system modeling as it emulates the full SoC elements. It can boot guest operating systems and emulate many ISAs, including ARMv7, ARMv8, and MicroBlaze. QEMU is a free and open source framework licensed under GPL-2.0.

For more information on QEMU, go to `https://www.qemu.org/`.

Xilinx has a QEMU port that's provided as a VP for SoCs built using the MicroBlaze processor, and both the Zynq-7000 and UltraScale+ SoC FPGAs. The platform can connect to custom IPs written in SystemC/TLM via an interface from the Xilinx QEMU.

For more information on Xilinx QEMU, check out its User Guide at `https://docs.xilinx.com/v/u/2020.1-English/ug1169-xilinx-qemu`.

Summary

This chapter opened *Part 2* of this book, which has a practical aspect to it since we will be putting the theoretical topics that were introduced in *Part 1* to use. This chapter was purely architectural since we need to understand why certain choices that we implement in an SoC design are the way they are. We also need to be capable of making certain changes to the design microarchitecture while considering the overall aspect of the system we are designing and whether we have met the stated objectives. This chapter covered all the major steps involved in SoC architecture design. We started by covering the exploration phase, where the possible design options are studied and compared in terms of cost, implementation effort, and time. We proposed a comparative method by which the initial theoretical analysis can be conducted and how the thinking process of choosing a potential solution can be driven. Then, we moved on to the next stage of the architecture definition, which was very analytical and was conducted practically on an example SoC, known as the ETS, which implements a low-latency dummy trading engine but behaves very much like one. We performed the hardware and software partitioning tasks on this trading engine by decomposing the SoC microarchitecture into many elements classified by tasks. While targeting the possible processing elements, we also looked at the end-to-end data path and what would make it low latency and easy to implant before putting together a table containing the choices we made. After that, we had to figure out how to make these processing elements communicate with each other by considering their specific characteristics. We listed these interfaces and how they should be dimensioned when such quantification is needed and makes sense. This collaboration between the software and hardware processing elements naturally led us to cover the Semi-Soft algorithm concept, where we covered its importance for SoC-based FPGA designs. We also covered many existing frameworks that are currently using it, starting from OpenCL, FPGA partial reconfiguration, and the simple hardware acceleration method we introduced in this chapter. We concluded this chapter by introducing the last stage of the architecture definition, known as system modeling. We covered how helpful it is nowadays for complex designs that specifically target ASIC technologies. We closed this chapter by providing the major frameworks currently used in the industry to perform system modeling and virtual prototyping.

In the next chapter, we will continue in the same vein by taking the ETS from its architecture definition to its implementation on the Zynq-7000 SoC FPGA. By doing so, you will learn how to translate the architecture choices into an implementable SoC for the ETS.

Questions

Answer the following questions to test your knowledge of this chapter:

1. Name some of the reasons that make an FPGA-based SoC the primary target technology for a full SoC system architecture implementation.

2. How does the architecture definition phase interact with the company's business strategy and its decision-making process?

3. List the main technical criteria that make a specific FPGA SoC a potential target device to implement the projected SoC architecture.

4. Summarize the methodology to use to start the architecture exploration phase, as presented in this chapter.

5. How is the architecture exploration phase concluded and what is its overall objective?

6. What step follows the architecture exploration phase and what is its main objective?

7. Describe the main functions that are performed by the ETS introduced in this chapter.

8. Why is UDP packet filtering chosen to be performed in hardware rather than in software?

9. How does the hardware-based UDP packet filtering performance compare to doing the same operation purely in software?

10. What is the main reason we choose to manage the Ethernet controller DMA descriptors in software using the pre-existing software drivers?

11. What kind of hardware-to-software and software-to-hardware communication and interfacing do we need to make hardware-based packet filtering work?

12. How did we make the UDP packet processing lower latency by introducing minimal disturbance to the software-only processing model? Why do we always aim to minimize the changes in any existing model?

13. Describe the Semi-Soft algorithm approach.

14. How was the Semi-Soft algorithm architecture design methodology used in the ETS?

15. List some of the main hardware acceleration techniques that are used at the system level and how they compare to using the Semi-Soft algorithm method.

16. What is OpenCL?

17. What is the purpose of system modeling and when is it performed?

18. What is a VP and why did we choose to use one?

19. List some of the major system modeling frameworks.

20. What are the operating modes of gem5?

7

FPGA SoC Hardware Design and Verification Flow

In this chapter, we will delve into implementing the SoC hardware of the **Electronic Trading System** (**ETS**), for which we developed the architecture in *Chapter 6, What Goes Where in a High-Speed SoC Design*. We will first go through the process of installing the Xilinx Vivado tools on a **Linux Virtual Machine** (**VM**). Then, we will define the SoC hardware microarchitecture to implement using the Xilinx Vivado tools. This chapter is purely hands-on, where you will build a simple but complete hardware SoC for a Xilinx FPGA. You will be guided at every step of the SoC hardware design, from the concept to the FPGA image generation. This will also cover hardware verification aspects, such as using the available RTL simulation tools to check the design and look for potential hardware issues.

In this chapter, we're going to cover the following main topics:

- Installing the Xilinx Vivado tools on a Linux VM

- Developing the SoC hardware microarchitecture

- Design capture of an FPGA SoC hardware subsystem

- Understanding the design constraints and PPA

- Verifying the FPGA SoC design using RTL simulation

- Implementing the FPGA SoC design and FPGA hardware image generation

Technical requirements

The GitHub repo for this title can be found here: https://github.com/PacktPublishing/ Architecting-and-Building-High-Speed-SoCs.

Code in Action videos for this chapter: http://bit.ly/3NRZCkU.

Installing the Vivado tools on a Linux VM

The Xilinx Vivado tools aren't supported on *Windows 10.0 Home* edition, so if you are using your home computer with this version installed on it, you won't be able to follow the practical parts of this book. Only *Windows 10.0 Enterprise* and *Professional* editions are officially supported by the *Vivado tools*. However, there are many Linux-based **Operating Systems (OSes)** that Xilinx officially supports.

One potential solution to build a complete learning environment using your home machine is to install the Vivado tools on a supported *Linux* version, such as *Ubuntu*, which you can run as a **VM** by using the *Oracle VirtualBox hypervisor* to host it.

The Vivado tools are officially supported on the following OSes:

- **Windows**: Windows Enterprise and Professional 10.0
- **RedHat Linux**: RHEL7, RHEL8, and CentOS 8
- **SUSE Linux**: EL 12.4, and SUSE EL 15.2
- **Ubuntu**: From 16.04.5 LTS up to 20.04.1 LTS

> **Information**
>
> For the full list of supported OS versions and revisions, please check out `https://www.xilinx.com/products/design-tools/vivado/vivado-ml.html#operating-system`.

For our practical design examples and to achieve the learning objectives, we can use a good specification home machine running Windows 10.0 Home edition as a learning platform. We will install Oracle VirtualBox on it, and then we will build a VM with *Ubuntu 20.04 LTS Linux* as a guest OS.

A good machine that we can use for the examples in this book should have at least the following hardware specification:

- **Processor**: Intel(R) Core(TM) i5-10400 CPU @ 2.90 GHz
- **Installed RAM**: 16.00 GB
- **System type**: A 64-bit OS and an x64-based processor

Installing Oracle VirtualBox and the Ubuntu Linux VM

Oracle VirtualBox is a hypervisor that can host the Linux VM. You can download it from `https://www.oracle.com/uk/virtualization/technologies/vm/downloads/virtualbox-downloads.html`.

You can install VirtualBox by double-clicking on the downloaded install executable – for example, we are using the `VirtualBox-6.1.34-150636-Win.exe` build.

You also need to download Ubuntu Linux from `https://ubuntu.com/#download`.

In VirtualBox, we can now proceed to build the Ubuntu VM by following these simple steps:

1. Go to **Machine** and select **New**, as shown here:

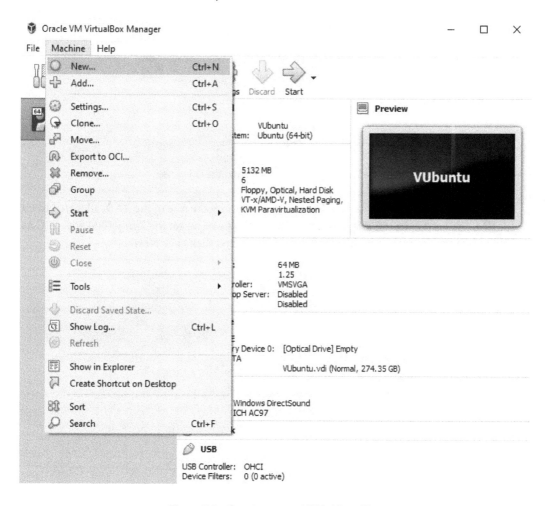

Figure 7.1 – Creating a new VM in VirtualBox

2. This will open the following window, where you can name the VM, for example, `UbuntuVM`, and then specify which guest OS we will be using (**Linux** and **Ubuntu (64-bit)**). Click on **Next**:

Figure 7.2 – Entering the name and the OS to use

3. This will open the following window, where we specify the size of the RAM allocated to the VM. We need at least 5 GB to run the Vivado tools and target the Zynq-7000 SoC FPGAs. Increase the value to something higher than 5 GB, and then click **Next**.

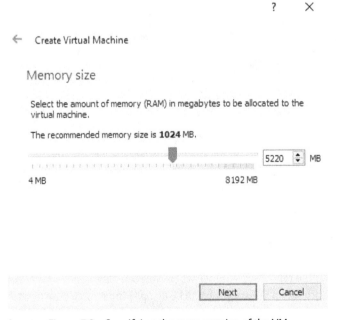

Figure 7.3 – Specifying the memory size of the VM

4. Now, we need to create a hard disk for the VM; choose the default and click on **Create**.

Figure 7.4 – Creating the hard disk for the VM

5. This will open the next window for specifying the type of virtual hard disk to create for the VM. Choose the default as shown and click on **Next**.

Figure 7.5 – Specifying the virtual hard disk type of the VM

6. Now, we need to specify the allocation of the storage of the virtual disk on the physical disk; choose an option from the two and then click on **Next**.

Figure 7.6 – Specifying the virtual hard disk storage allocation type

7. Now, we need to specify the size of the virtual hard disk, as shown in the following figure. Choose something around 200 GB if you have spare space on your computer, as we need space for the Vivado tools (50 GB) and space for the installation process. Click on **Create**.

Figure 7.7 – Specifying the size of the virtual hard disk

The VM is now created, as shown in the following figure.

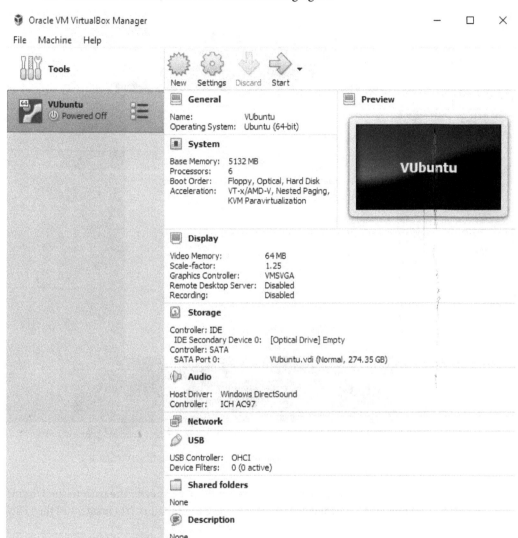

Figure 7.8 – The UbuntuVM VM created in VirtualBox

8. To start the VM and install the Ubuntu Linux OS on it, just double-click on it or use the green arrow on the right-hand side. This will open the following window. The first time this VM boots, we need to specify the OS image via the window that opens automatically. Click on the green arrow in the folder icon.

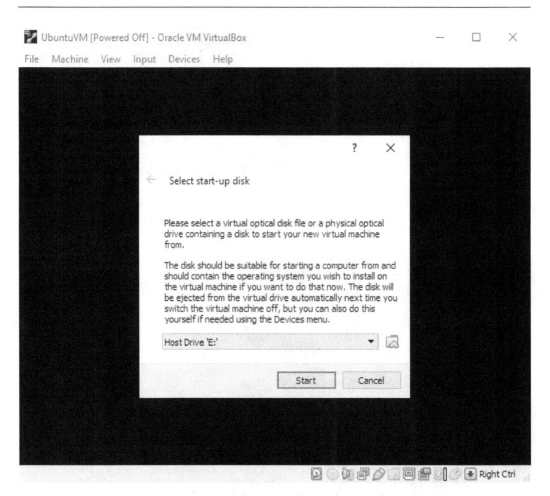

Figure 7.9 – Launching the UbuntuVM VM for the first time from VirtualBox

9. The following window will open. Click on the **Add** sign to specify the path to the Ubuntu Linux ISO file previously downloaded. Browse to the Ubuntu Linux ISO image, and then click on **Choose**.

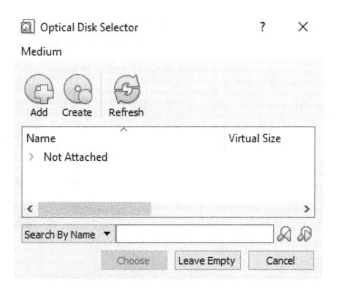

Figure 7.10 – Specifying the OS to use with the VM

10. This will open the launch window of the Ubuntu VM, as shown here. Click on **Start**.

Figure 7.11 – Starting the UbuntuVM VM

11. In the screen that follows, just click on *Enter* to install Ubuntu Linux on the VM as the guest OS. Once the Linux OS boots, just follow the instructions to set it up for the first time.

We now have an officially supported OS that can be used for our learning objectives.

Installing Vivado on the Ubuntu Linux VM

From within the Ubuntu Linux VM, download the Xilinx Vivado installer for Linux from `https://www.xilinx.com/support/download.html`.

To gain access and be able to download Vivado, you will need to register first.

Before starting the installation process of Vivado, we need to make sure that we have all the packages required by Vivado to install and operate correctly on Ubuntu Linux. We found that the following libraries and packages are missing in the default Ubuntu Linux installation and simply need to be added:

- `libtinfo5`
- `libncurses5`
- `"build-essential"`, which includes the GNU Compiler Collection

To install the preceding packages, simply use the following command from within a command shell in Ubuntu:

```
$ sudo apt install libtinfo5
```

Then, for `libncurses5`, use the following command:

```
$ sudo apt install libncurses5
```

And for the `build-essential` package, use the following command:

```
$ sudo apt update
$ sudo apt install build-essential
$ sudo apt-get install manpages-dev
```

We should have an Ubuntu Linux build ready to install the Vivado tools.

Now, we need to launch the installer of the Vivado package. From Command Prompt, use `cd` to go to the *download location* and change the *binary* file attribute to be executable:

```
$ sudo chmod +x Xilinx_Unified_2022.1_0420_0327_Lin64.bin
```

Then, launch it using the following:

```
$ sudo ./Xilinx_Unified_2022.1_0420_0327_Lin64.bin
```

This will start the installation process by first downloading the necessary packages and installing them. This process may take some time, given the size of the Vivado software package, and how long the installation process will last will depend on your internet connection speed. It is typical to leave it overnight, even over a fiber optic internet connection.

Once Vivado has finished installing, we suggest changing the settings of the Vivado IDE source files Text Editor that is enabled by default in Vivado from **Sigasi** to **Vivado**. This will avoid an issue using this Text Editor. When launching any command that requires using the preceding text editor selected by default, we noticed on the UbuntuVM Linux VM that it causes Vivado to get stuck when trying to launch the text editor. To change the Text Editor, simply follow these steps:

1. Go to **Tools**.
2. Select **Settings**.
3. Select **Tool Settings**.
4. Then, select **Text Editor**.
5. Then, select **Syntax Checking**.
6. Change the value of the **Syntax Checking** field from **Sigasi** to **Vivado**.
7. Restart Vivado.

Developing the SoC hardware microarchitecture

In the previous chapter, we defined the SoC architecture and left it to the implementation stage to choose the optimal solution that meets the specification of the low-latency **ETS**. We also left the exact details of the **Electronic Trading Market Protocol (ETMP)** to be defined here as close to the implementation stage as possible, just for practical reasons, since this protocol is really defined by the **Electronic Trading Market (ETM)** organization. We will define a simple protocol that uses UDP as its transport mechanism between the ETM network and the ETS. We have also not defined the exact details of the communication protocol between the software and the hardware PE for sending the requests to filter from the Cortex-A9, processing them by the hardware PE, and posting the results back from the hardware PE to the Cortex-A9. This protocol will be defined here by finding a fast exchange mechanism to be used by the microarchitecture proposal.

The ETS SoC hardware microarchitecture

From the ETS SoC architecture specification, we can draw a proposal SoC hardware microarchitecture that can implement the required low-latency ETS SoC.

Defining the ETS SoC microarchitecture

We need to build the following functionalities in the hardware acceleration PE:

1. Listen to the software via a doorbell register for requests to filter the received Ethernet packets on the Ethernet port:

    -   On receiving the Ethernet packets via its DMA engine, the software prepares a request for acceleration by providing an **Acceleration Request Entry** (**ARE**) in the **Acceleration Request Queue** (**ARQ**), located in local memory within the **Programmable Logic** (**PL**) side of the SoC.

    -   This request is in a form of an entry that has three fields: the address of the first DMA descriptor, the number of received packets, and an **Acceleration Done** status field.

    -   The length of the ARQ circular buffer needs to be computed according to the speed at which the UDP frames are received from the Ethernet port, and the speed at which they are processed by the hardware accelerator PE.

    -   The Cortex-A9 should never block on a full request circular buffer.

2. The software notifies the hardware accelerator engine that there are received Ethernet frames within the memory that need filtering and processing.

3. The hardware accelerator engine receives the notification and information from the Cortex-A9 and performs the packet processing of the Ethernet frames that are found to be UDP packets.

4. The hardware accelerator populates the urgent buying queue and notifies the trading algorithm task via an interrupt when it finds a matching symbol.

5. The hardware accelerator puts the address (pointer) of the DMA descriptor associated with the UDP packets in the **DMA Descriptors Recycling Queue** (**DDRQ**) for the **DMA Descriptors Recycling Task** (**DDRT**) to recycle them.

6. The hardware accelerator populates the urgent selling queue and notifies the trading algorithm task via an interrupt when it finds a matching symbol.

7. The hardware accelerator puts the DMA descriptor associated with the UDP packets in the DDRQ for the DDRT to recycle them.

8. The hardware accelerator populates the **Market Data Queue** (**MDQ**) and sends a notification to the **Market Database Manager Task** (**MDMT**) via an interrupt when it finds a matching symbol.

9. Once the Ethernet frames have all been inspected, the hardware accelerator notifies the Cortex-A9 via an interrupt that it has processed all the Ethernet frames from the last ones it has been asked to deal with.

10. The Cortex-A9 checks the received Ethernet frames for the ones left untreated by the hardware accelerator (non-UDP packets). When it finds some, it consumes them. The Cortex-A9 software can use any scheme to keep track of these among the ones that were accelerated by the hardware PE.

11. The Cortex-A9 changes the value of the **Ownership** field in the DMA descriptors for the Ethernet frames that it dealt with itself.

12. The Cortex-A9 notifies the Ethernet drivers or the DDRT to perform any DMA descriptor recycling actions.

From the preceding request-response summary between the Cortex-A9 and the hardware acceleration PE and the SoC architecture specification, we can come up with many low-latency microarchitecture proposals; one of the simplest designs for the implementation is depicted in the following diagram:

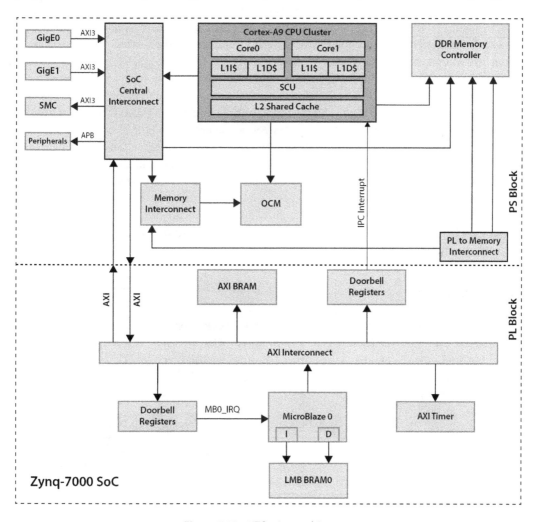

Figure 7.12 – ETS microarchitecture

This microarchitecture is simple to implement, as it uses the MicroBlaze processor as the processing engine of the hardware acceleration. It is well suited for performing packet processing, as it allows the experimentation of many parallel processing algorithms. We can define the best approach that suits the ETS even post-deployment, since this is a real semi-soft algorithm implementation. If we decide to redesign the hardware accelerator as **Multiple Data Multiple Engines** (**MDMEs**), we can do so by using multiple MicroBlaze processors and dispatching filtering and packet processing work fairly among them. This will require triplicating the Cortex-A9 interfaces listed previously in this subsection. This is also a scalable microarchitecture, meaning that if we judge that we need a higher processing rate, we can then augment the number of MicroBlaze processors, or increase the operating clock frequency (while the FPGA implementation permits), and obviously reimplement the design. We will also need to reconfigure the FPGA as well as reimplement the Cortex-A9 software, but it can all be done, and in the least disruptive manner, if the FPGA still has the necessary resources to implement extra MicroBlaze-based CPUs. If not, this may require designing newer hardware using the next density available FPGA SoC, using the same package from the Zynq-7000 FPGAs. The dispatch model could also operate in a **Single Data Multiple Engines** (**SDMEs**) model. In the SDME model, and for a faster match lookup, the Cortex-A9 can request that all the MicroBlaze processors concurrently search for a specific filter match each, but on the same Ethernet packet. This may speed up the filtering by executing fewer branches on the MicroBlaze software but will increase the complexity of all the software. In this compute model, access to shared data should be atomic among the different MicroBlaze processors as the packet filtering becomes more distributed. However, this is perfectly fine as another microarchitecture proposal option, which will then also be retained. We suggest using the MDME dispatch model for the hardware acceleration engine microarchitecture, and we will be implementing the necessary queues and notification mechanisms for it to work in the next chapter, when we start implementing the ETS SoC software.

ETMP definition

We will define a simple ETMP to transport over a UDP packet for the ETM, for both the **Market Management** (**MM**) and the **Market Data Information** (**MDI**) packets. For the trading TCP/IP, we will leave it until we start developing the Cortex-A9 software in *Chapter 8, FPGA SoC Software Design Flow*. The ETMP defines a single-length UDP packet payload of 320 bits/40 bytes and has many fields, as defined in the following table:

Field	Length in bits	Description
Symbol Code (SC)	32	The financial traded product symbol code. Every financial product has a unique code assigned by the ETM when the product is first introduced to the ETM.
Packet Type (PT)	32	States whether this is an MM packet or a Market Data packet: 0b0: Market Data packet 0b1: Management packet

Field	Length in bits	Description
Proposed Volume (PV)	32	The proposed maximum volume for a sell or buy action. Partial proposals of trade can be made by the ETS if interested in the symbol.
Transaction Type (TT)	32	The transaction type associated with this financial product, buying or selling: 0b0: Buying 0b1: Selling
Timestamp (TS)	64	This is the timestamp logging from when the UDP packet left the ETM servers.
Day (D)	32	Encodes the day when the UDP packet was sent.
Month (M)	32	Encodes the month when the UDP packet was sent.
Year (Y)	32	Encodes the year when the UDP packet was sent.
Error Detection Code (EDC)	32	CRC32 computed over all the ETMP packets, excluding itself (over 288 bits)

Table 7.1 – The ETMP packet format

The Full ETMP UDP packet within the UDP frame is depicted in the following figure:

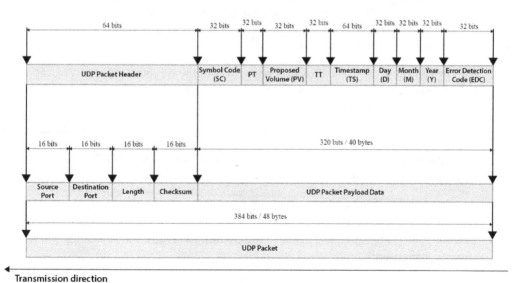

Figure 7.13 – The ETMP packet layout

The UDP header adds another 64 bits of data to the packet, resulting in an ETMP UDP frame of 384 bits/48 bytes, as illustrated in the preceding figure.

There is a possibility to implement a CRC32 calculator in the hardware and connect an instance of it to the MicroBlaze processor, to use for confirming the integrity of the ETMP packet. The alternative solution would be to start with a software function to perform it and deal with it in hardware post-profiling. There are open source implementations of the CRC32 algorithm that we can use, or we can build our own CRC32 function in the software. All of these are options we can decide upon by running a benchmarking exercise on a MicroBlaze-based hardware design, where we profile the software-based CRC32 algorithm implementation and assess its suitability for the low-latency ETS. We will revisit this topic as part of the hardware acceleration in *Part 3* of this book.

Design capture of an FPGA SoC hardware subsystem

In this section, we start the building process of the ETS SoC hardware subsystem using the Xilinx Vivado tools. We will start by creating a Vivado project, adding the required subsystem IPs, configuring them, and connecting them to form the SoC using the **IP Integrator** utility of Vivado. But first, we will create a Vivado project that targets one of the Zynq-7000 SoC demo boards, if we have it at hand; we can then use it to verify the final functionality of the ETS SoC once we have built software for it. Also, any available demo board capable of hosting a Zynq-7000-based SoC design can be used as a target. These design capture steps were introduced in *Chapter 2, FPGA Devices and SoC Design Tools,* of this book, and we will build upon this information to achieve our current objective.

Creating the Vivado project for the ETS SoC

The first step is to launch the Vivado GUI:

1. Start by launching the VirtualBox hypervisor to boot the UbuntuVM Linux VM. If you are using Windows Enterprise or Professional as a host machine, it is still recommended to use a Linux VM for this book's projects.

2. Now, we need to launch the Vivado GUI by using Command Prompt and typing the following:

```
$ cd <Tools_Install_Directory>/Xilinx/Vivado/2022.1/bin/
$ sudo ./vivado
```

Replace <Tools_Install_Directory> with the path where you have installed Vivado on your UbuntuVM Linux VM. When prompted for the sudo password, provide the root password you set up when you installed the UbuntuVM Linux VM (assuming you haven't changed it afterward).

Once Vivado is up and running, we can then create the ETS SoC project by doing either of the following:

- Starting from an existing project template and customizing it to our needs

- Creating the project from scratch, using IP Integrator to add components, and then configuring them

Since we have already gone through the creation of an empty Vivado project in *Chapter 2, FPGA Devices and SoC Design Tools*, we can now start our design from an existing project template; we will then customize it to our microarchitecture needs and add the necessary IPs, using IP Integrator to implement the hardware PE for the packet filtering tasks. This is a better approach when you are just starting the learning journey of the Xilinx FPGA SoC designs and the hardware design flow. This promotes design reuse, minimizes design bugs, and speeds up the integration process. Let's begin.

3. In the Vivado **Quick Start** window, click on **Open Example Project**. This will launch the following menu:

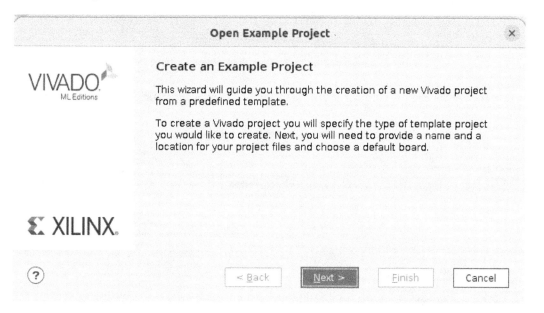

Figure 7.14 – Creating a Vivado project, starting from a predefined template

4. Click **Next >** in the preceding window; this will then open the **Select Project Template** window. Choose the **Zynq-7000 Design Presets** option in the **Templates** section, and then click on **Next >** once you have read its description.

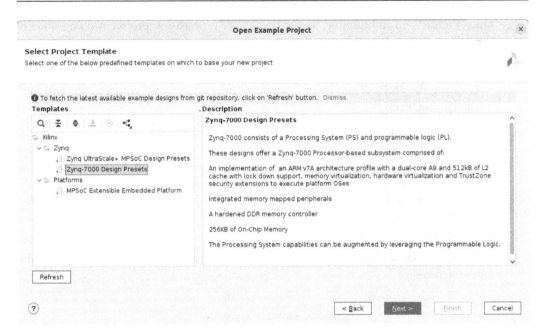

Figure 7.15 – Selecting the predefined template for the ETS SoC

5.   Specify the ETS SoC project details, as shown in the following figure, and then click **Next >**.

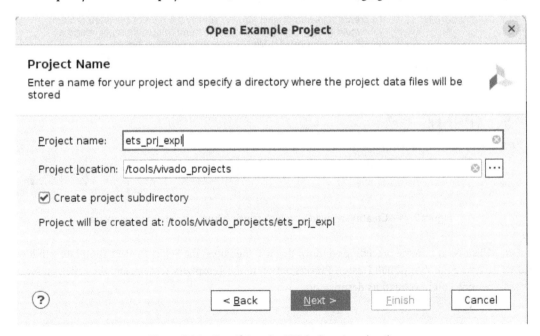

Figure 7.16 – Specifying the ETS SoC project details

6. This will open the following window to select a default Xilinx board for this project. Select the board as highlighted in the following screenshot, and then click **Next >**.

Figure 7.17 – Choosing the default board for the ETS SoC project

7. The selection of a design preset can be performed in the following window, as indicated by the following figure, to include both the PS and PL in the ETS SoC. Once selected, click **Next >**.

Figure 7.18 – Choosing the design preset for the ETS SoC project

8. The last figure in this project template selection is a summary of the chosen preset for the ETS SoC. Click on **Finish** to open the selected preset project.

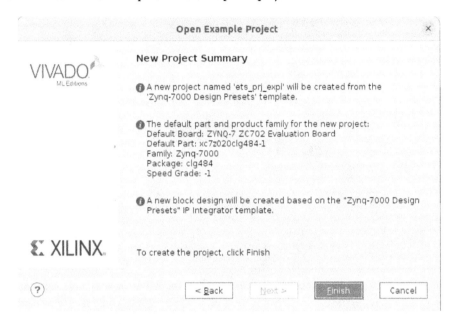

Figure 7.19 – The selected design preset summary for the ETS SoC project

9. The ETS SoC project will open, as shown in the following figure. We can now start to explore what is included in it so that we can extend it to meet the ETS SoC hardware microarchitecture needs.

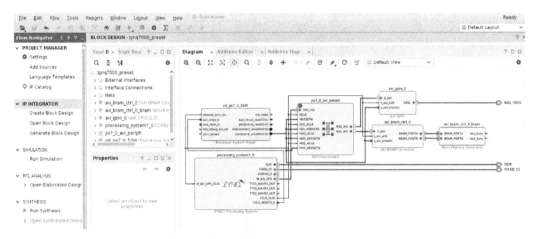

Figure 7.20 – The Vivado ETS SoC project starting preset view

Revisiting the ETS SoC microarchitecture and looking at the preceding Vivado project, we have both the PS and the PL blocks of the FPGA included. In the PL block, we have an AXI BRAM, an AXI GPIO, and an AXI interconnect. We shall keep all these elements, and we shall add the MicroBlaze processor and its required peripherals. For the Cortex-A9 to the MicroBlaze processors' **Inter-Process Communication (IPC)** interrupt, we will simply use an AXI **Interrupt Controller (INTC)** IP. The INTC will implement the required doorbell registers, which have the capability to generate software-triggered interrupts. We only need to add these later using IP Integrator, configure them, and then revisit all the preset IPs in the PL block from the template, which should still be valid; if not, correct their settings. Before moving on to the PL block subsystem design task, let's first make sure the preset configuration of the PS block is in line with our desired features for the ETS SoC microarchitecture. To do so, we can graphically examine the settings and adjust them as needed.

Configuring the PS block for the ETS SoC

In the **IP Integrator** main window, double-click on the Zynq processing system; this will open the following configuration window. All the parts that are highlighted in green are user-customizable. Let's proceed to the PS customization task:

1. Let's unselect the peripherals we don't need for the ETS SoC, such as **CAN 0**, **SD0**, and **USB 0**. To disable these peripherals in the PS, click anywhere in the I/O peripherals.

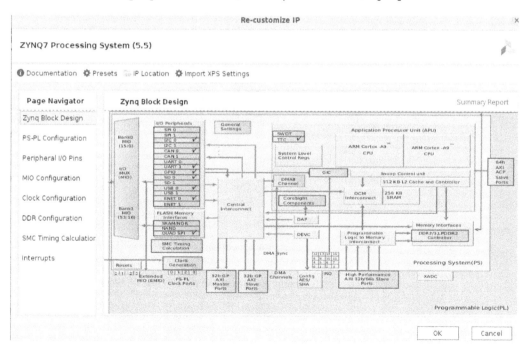

Figure 7.21 – Customizing the peripherals to use in the ETS SoC PS block

2. This will open the I/O peripherals selection window. Uncheck the unwanted peripherals, as shown here.

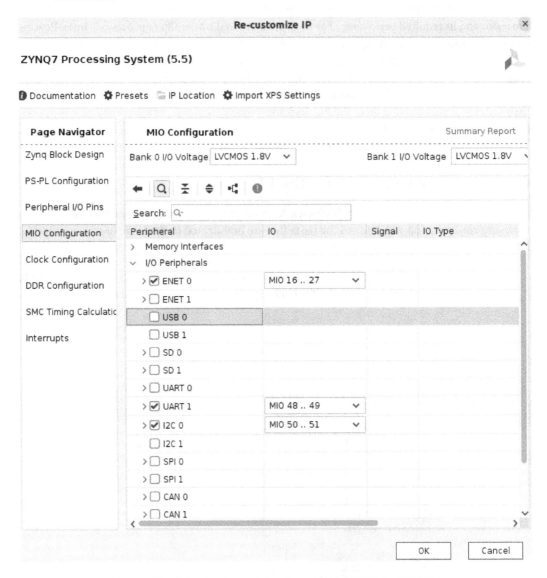

Figure 7.22 – Selecting the peripherals to use in the ETS SoC PS block

3. Now, let's move to the next customization step by selecting the **Peripheral I/O Pins** row in the **Page Navigator** pane on the left-hand side of the window; this will open the following view, showing the current assignment of the selected peripherals in the PS block and how they are connected to the external world – that is, via the MIO path and not the EMIO path. We introduced these connectivity options in *Chapter 2, FPGA Devices and SoC Design Tools*. We should keep the chosen option to match the selected board layout; if we have another alternative board design, we can route them when applicable through the EMIO path via the FPGA PL block.

Figure 7.23 – Customizing the ETS SoC PS peripheral I/O pins

4. Select the **MIO Configuration** row, which will open the **MIO Configuration** window, showing the selected pins for every I/O of every peripheral selected in the PS block.

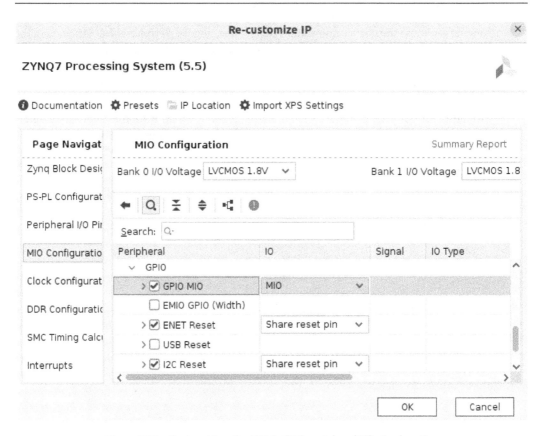

Figure 7.24 – Customizing the ETS SoC PS peripheral I/O pins location

5. Select the next row down, which is the **Clock Configuration** row. We can customize the clocking on the SoC in this window. Set the values as shown in the next figure, setting up the CPU, the DDR, and the PL block interface clocking.

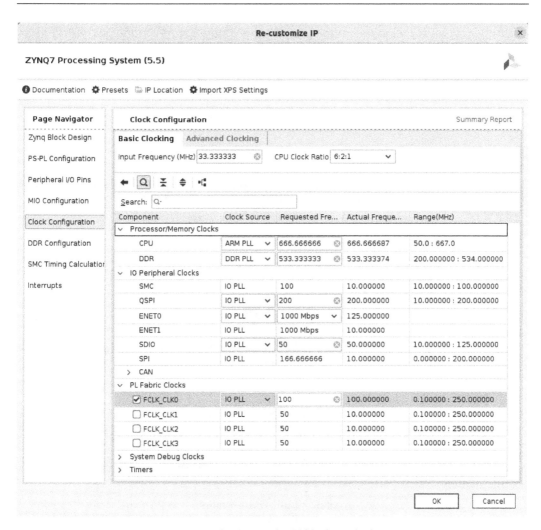

Figure 7.25 – Setting up the PS block IPs clocking

6. Select the **DDR Configuration** row to explore all the available controllers, devices, and board parameters. Leave everything as it is, since it is set for the selected demo board. Anything that needs to be specifically configured for the ETS SoC can be done via the DDR controller registers visible to the software.

7. As shown in the following figure, select the **Interrupts** row to examine the available customization options. Expand the **PL-PS Interrupt Ports** row, and then add **IRQ_F2P[15:0]** to enable the 16-bit shared interrupts from the PL. As indicated, these are wired to the Cortex-A9 **Generic Interrupt Controller** (GIC), and they are connected to the **Shared Peripheral Interrupt** (SPI) inputs, positioned at **[91:84]** and **[68:61]**.

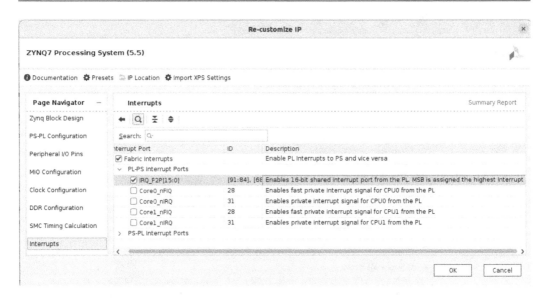

Figure 7.26 – Customizing the SoC interrupts

8.   Once done, click on **OK**. Now, let's make sure that the customization thus far is correct. To do so, go to the Vivado IDE main menu and click on **Tools | Validate Design**. This will come up with no error.

The preceding steps conclude the PS customization of the generated design template. The next subsection will focus on the PL block and the required ETS SoC microarchitecture IPs, configuration, and connectivity.

> **Information**
>
> Further practical information on using Vivado IP Integrator to design a SoC using the Zynq-7000 PS block is available in *Chapter 3* of *Vivado Design Suite User Guide: Embedded Processor Hardware Design* at `https://docs.xilinx.com/v/u/2021.2-English/ug898-vivado-embedded-design`.

Adding and configuring the required IPs in the PL block for the ETS SoC

As already mentioned, we need to design a hardware accelerator for the ETM UDP packet filtering, which will use a MicroBlaze-based **Packet Processor** (**PP**) subsystem. The PP will have an associated **AXI Interrupt Controller** (**AXI INTC**) to connect any interrupting IPs and implement the necessary IPC interrupt from the Cortex-A9 processor. The AXI INTC supports SW-generated interrupts where the Cortex-A9 writes to an AXI INTC internal register to interrupt the MicroBlaze processor. We will also add an AXI Timer and another AXI INTC to implement the IPC interrupt in the opposite direction – that is, from the MicroBlaze processor toward the Cortex-A9 processor. We will use IP Integrator to design this part of the ETS SoC.

> **Information**
>
> If you have closed the Vivado GUI and are reopening the tools to continue the design process, then once you launch Vivado, the list of available projects is displayed on the right-hand side of the launch screen. To open one, simply click on the project name.

Once the ETS SoC project, `ets_prj_exp1`, is reopened in Vivado, to restore the **IP Integrator** view, just click on **Open Block Diagram** under the **IP Integrator** row in the **Flow Navigator** pane of Vivado, located on the left-hand side menu in the GUI. Now, we will add the MicroBlaze processor subsystem to implement the PP design:

1. Click on the + sign at the top of the **IP Integrator** window. The IP selection filter will open. Type `microblaze` into the filter field, and then you will see the relevant IPs displayed, as shown in the following figure. Select **MicroBlaze** from the list and hit *Enter*.

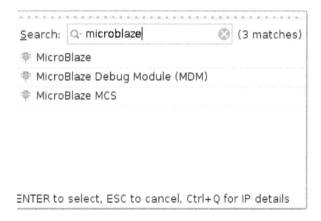

Figure 7.27 – Searching for the MicroBlaze IP in the IP Integrator

2. Once the MicroBlaze IP has been added, click on the **Run Block Automation** *hyperlink* that appears in the **IP Integrator** main window, as shown here.

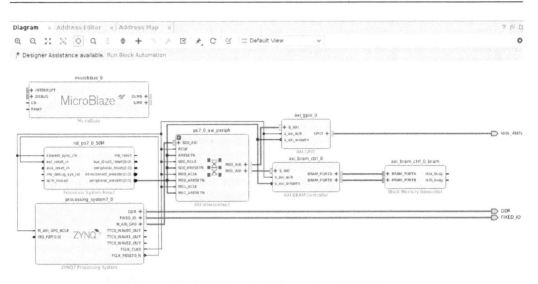

Figure 7.28 – Launching Run Block Automation in the IP Integrator

3. This will open the MicroBlaze subsystem *configuration wizard*. Set its configuration as shown here and click **OK**.

Figure 7.29 – Configuring the MicroBlaze subsystem in the IP Integrator

4. Once the MicroBlaze processor is configured, remove the *MB AXI interconnect* from the PL subsystem by right-clicking on it and then selecting **Delete**. We will use a single AXI interconnect in the PL block to network both the MicroBlaze subsystem and the PS.

5. We need to customize the MicroBlaze-embedded processor subsystem. In the **IP Integrator** main window, double-click on the MicroBlaze instance added to the design. This will open the MicroBlaze customization window, as shown in the following screenshot. Select **Microcontroller Preset** in the **Select Configuration** field. Make sure that the other selected settings match the figure. Then, click on **Next >**.

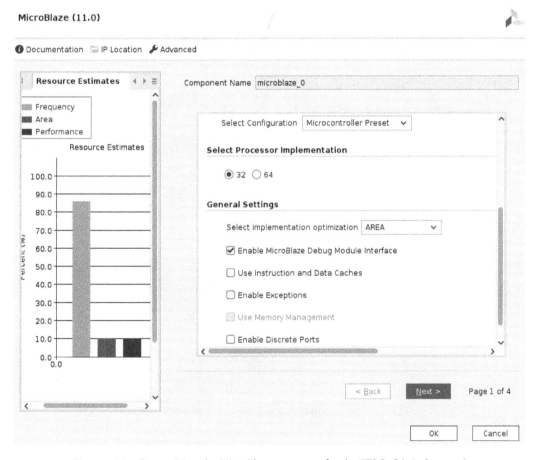

Figure 7.30 – Customizing the MicroBlaze processor for the ETS SoC (window one)

6. This will open the next MicroBlaze processor configuration window. In the **Optimization** section, select **PERFORMANCE** and set all the other parameters, as shown in the following figure. Then, click **Next >**.

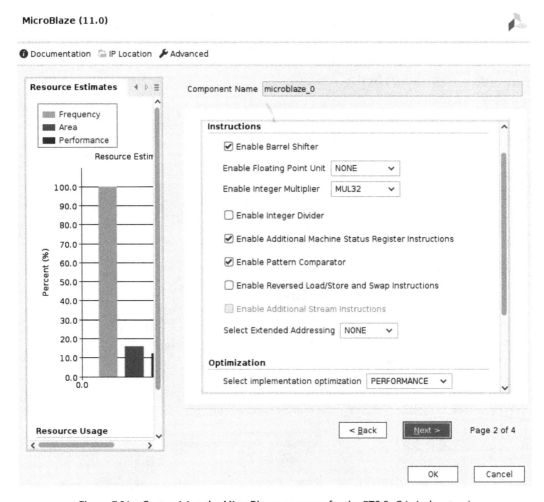

Figure 7.31 – Customizing the MicroBlaze processor for the ETS SoC (window two)

7. In the third MicroBlaze IP configuration window, select **BASIC** as the **Debug** mode and leave the other fields as their defaults. Then, click **Next >**.

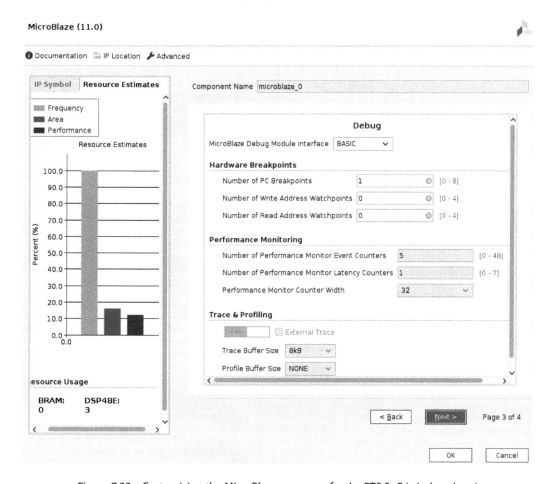

Figure 7.32 – Customizing the MicroBlaze processor for the ETS SoC (window three)

8. Once the fourth MicroBlaze IP configuration window is open, make sure that the **Local Memory Bus (LMB)** is enabled for both the instruction and data sides of the processor. We intend to run the MicroBlaze-embedded software from the LMB memory, which provides local SRAM with fast access. Make sure that all the other settings match the following figure. Then, click on **OK**.

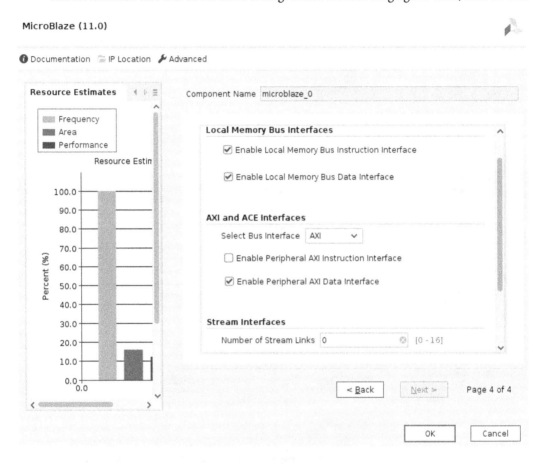

Figure 7.33 – Customizing the MicroBlaze processor for the ETS SoC (window four)

9. Now, double-click on the PS AXI interconnect (`ps7_0_axi_periph`) and configure it for the desired number of slave and master ports, as shown in the following figure. Then, click **OK**.

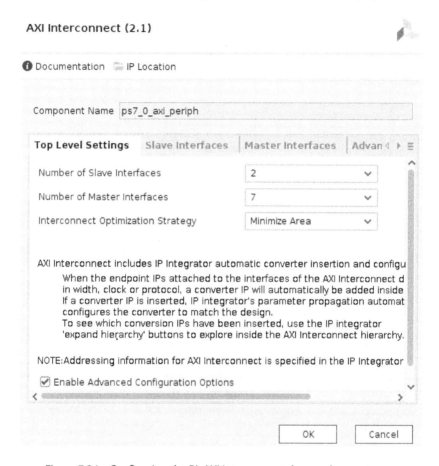

Figure 7.34 – Configuring the PL AXI interconnect slave and master ports

---

Information

Make sure that the **Interconnect Optimization Strategy** option in the preceding **AXI Interconnect** configuration window is set to **Minimize Area**. This will prevent hitting a known issue with Vivado IP Integrator, which should come up with an error code, **[BD 41-237] Bus interface property ID_WIDTH does not match**, if **Interconnect Optimization Strategy** is not set to **Minimize Area**.

10. Connect the MB AXI Data Master (M_AXI_DP) port to ps7_0_axi_periph, and do the same for all the MicroBlaze peripherals. The resulting system should look like the following figure.

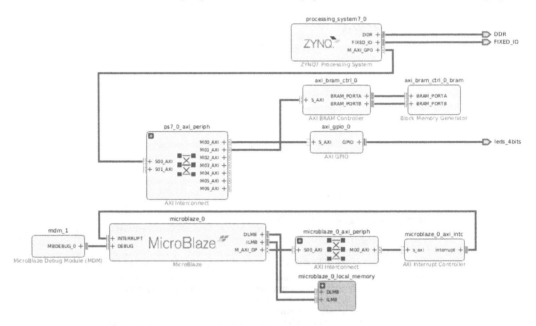

Figure 7.35 – The PL subsystem view following the customization

11. Now, we will add the AXI Timer, and then click on the **Run Connection Automation** command suggested by **Designer Assistance** in IP Integrator. We will then complete the process manually and connect the AXI Timer interrupt output to the MicroBlaze AXI INTC (microblaze_0_axi_intc) via the concatenation vector (xl_concat_0). The second entry of the concatenation vector is from the **MicroBlaze Debug Modem** (MDM), as highlighted in the following figure.

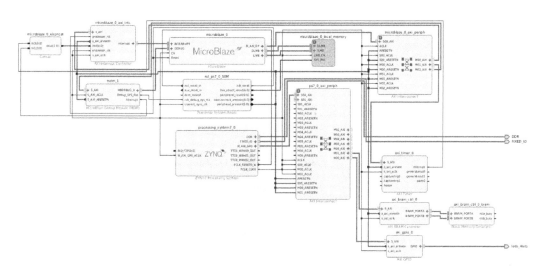

Figure 7.36 – MicroBlaze processor subsystem interrupts connectivity

12. To implement the IPC interrupt from the MicroBlaze to the Cortex-A9, we can use a second AXI INTC instance. Add an AXI INTC using the IP Integrator. At the *AXI INTC IP* configuration stage, select **8** for **Software Interrupts** in the **Advanced** configuration tab. The following figure shows the desired configuration for the second *AXI INTC IP*.

Figure 7.37 – Adding an AXI INTC for the IPC interrupt from the MicroBlaze to the Cortex-A9

13. Now, we need to connect the second AXI INTC `axi_intc_0` interrupt output to the `IRQ_F2P` interrupt input of the PS block that we enabled previously when customizing the PS block. We simply need to right-click on the `axi_intc_0` interrupt output port, which will bring up the menu as shown here. Select **Make Connection….**

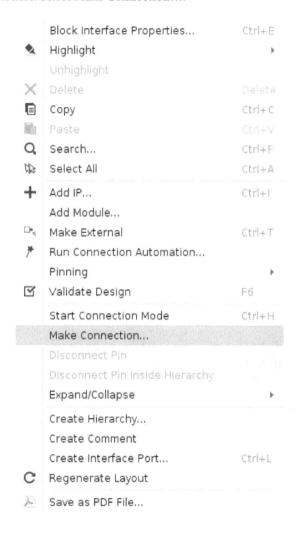

Figure 7.38 – The IP Integrator signal connectivity feature

14. The following figure will open with the possible connection options for the AXI INTC interrupt output signal. Select the desired port, IRQ_F2P, and click **OK**.

Figure 7.39 – Connecting the AXI INTC interrupt output to the IRQ_F2P PS input

15. We should have all the IPs required to build the PP, based on a MicroBlaze processor that is included, configured, and connected using the IP Integrator GUI. The resulting subsystem should look like the following figure.

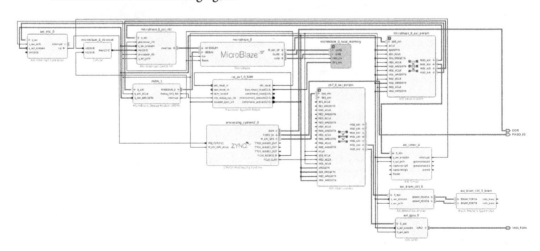

Figure 7.40 – The PS and PL subsystem view following the customization

16. We should now proceed to the subsystem address mapping by selecting the **Address Map** tab in the **IP Integrator** main window, which will open the system **Address Map** view. This needs to be set up to match the customization desired in the PP subsystem. There are unassigned regions and excluded regions by default. To include more IPs as visible regions in the MicroBlaze address map or the PS master address map, first expand the MicroBlaze address map region, select the IP or IPs to add, right-click on the IPs, and click on **Assign**, as shown in the following figure.

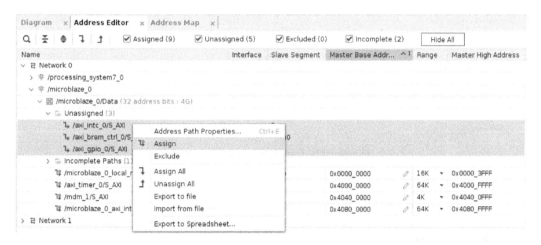

Figure 7.41 – Assigning IPs to the MicroBlaze data side address map region

17. Once an IP has been assigned, it now needs to be included in the address map by right-clicking on the IP row and then clicking on **Include**.

18. The MicroBlaze data address map should look like the following:

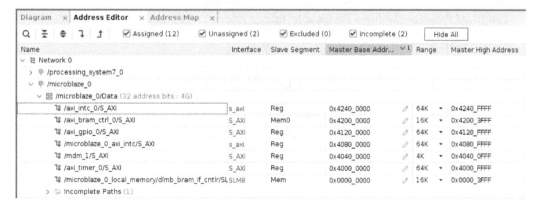

Figure 7.42 – The MicroBlaze data side address map

19. The PS master visible address map within the PL block needs to be configured to look like the following figure. You can use the same techniques described for the MicroBlaze address map customization to set it up.

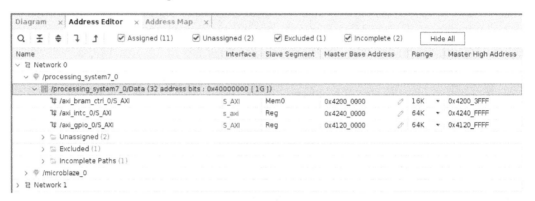

Figure 7.43 – The PS master address map

20. We can increase the MicroBlaze *LMB* size to 16 KB in anticipation of the MicroBlaze software being more than the allocated 8 KB in the preset. Expand **Network 1** in the preceding figure and then set the *LMB* size to 16 KB. Click on the **Diagram** tab of IP Integrator to return to the system view.

21. Go to the Vivado IDE main menu and click on **Tools | Validate Design**. This should return no errors on the ETS SoC design.

> **Information**
>
> Further practical information on using Vivado IP Integrator to design a MicroBlaze-based embedded processor subsystem is available in *Chapter 4* of *Vivado Design Suite User Guide: Embedded Processor Hardware Design* at https://docs.xilinx.com/v/u/2021.2-English/ug898-vivado-embedded-design.

Understanding the design constraints and PPA

This section will briefly cover Power, Performance, and Area (PPA) analysis and the physical design constraints required when the SoC RTL is implemented by the tools targeting the FPGA technology.

What is the PPA?

PPA is an analysis we usually perform on a given target IP when designing an SoC for the ASIC technology. This is performed to evaluate its characteristics under these metrics. This analysis provides us with an idea of the IP requirements before the IP is physically implemented in the actual ASIC device. The study helps us in understanding the following:

- The IP consumption in terms of power, measured in watts. The static power is consumed by simply powering up the IP and the dynamic power when the IP executes the type of workload it is supposed to perform.

- The maximum clock frequency measured in MHz at which the IP can run, which will give us an idea of whether the IP meets the system requirement in terms of performing the tasks at the speed needed.

- The IP size in terms of the silicon area needed to implement it within the target SoC, measured in μm^2 to perform the required functionality. This also includes any storage it may require, such as **SRAM** and **Read-Only Memory (ROM)**.

Although for the FPGA these metrics can be measured by simply implementing the design (or a cut-down version of it to only include the IP on its own), when the design or the IP is available, we still need to know these values prior to the implementation trials. These figures help in dimensioning the FPGA capacity, its speed grade, and the supporting circuits to use on the electronics board, such as the power supplies and external storage.

Also, when selecting an FPGA for our SoC, many choices of the implementation decision are already implicitly made by the FPGA technology vendor. This means that we are in the process of choosing the optimal device that can host our SoC; usually, we should oversize by some margin for future expansion, just in case there is a need to add a feature or adjust an IP capability once deployed in the field. This is an FPGA, which means that it can be reprogrammed or patched at any moment, and that can be after the solution deployment.

We will briefly cover this subject in this chapter to provide an overview of what PPA means for FPGA designs. You are encouraged to research the topic further to make physical design decisions that may affect the overall success of your project. When implementing an SoC following all the design stages covered thus far in *Part 2* of this book, we need to understand that a given IP needs resources to be implemented and mapped to the FPGA available logic.

> **Information**
>
> You are encouraged to study the PPA topic further if you are not familiar with these metrics; ARM provides a good tutorial on the subject that can be accessed at `https://developer.arm.com/documentation/102738/0100/?lang=en`.

Synthesis tool parameters affecting the PPA

The way an IP is synthesized and implemented by tools may affect the target IP PPA. If the logic elements to which the RTL has been mapped are geographically located closer to each other and use the fastest available nets for interconnections, the result may be a fast IP able to run at high frequencies. Sometimes, we need to make compromises between design speed and the required space within the FPGA. Also, the larger a design is in terms of silicon area, the more static and dynamic power it requires to deliver the performance we are after. Many other interdependencies need to be understood before setting up the required constraints at all the design stages, in order to guide the implementation tools to achieve our design implementation objectives. The synthesis tools have many techniques at hand to try and achieve a design objective, such as the following:

- Register replication to augment the fan-out or reduce the net delays of the connecting signals of two logic elements that are geographically far apart in the silicon area.

- **Finite State Machine** (**FSM**) encoding, which affects the area and the speed at which FSM can run.

- Merging and optimizing logic by eliminating duplicate logic, such as registers used for synchronization during the input or output of an IP, which when connected may end up using both. However, sometimes we want this to be the case.

There are many options and attributes that we can specify for the synthesis tool to guide its optimization strategy and, therefore, affect the PPA outcome. Usually, there are profiles for speed, area, or a balanced strategy that presets all the synthesis parameters to a default value without the requirement to set each one of them. It is good to familiarize yourself with these to understand how these parameters affect the outcome of the synthesis stage and, if needed, to tune them to produce a different or better outcome. *Chapter 2* of *Vivado Design Suite User Guide: Synthesis* provides a detailed list of all the available attributes that a user can set at this first stage of the FPGA design implementation, which can be found at `https://www.xilinx.com/content/dam/xilinx/support/documents/sw_manuals/xilinx2022_1/ug901-vivado-synthesis.pdf`.

Specifying the synthesis options for the ETS SoC design

We first need to generate the design RTL files for the ETS SoC project using the Vivado GUI, as indicated in *step 1*; then, we can proceed to specify the project synthesis constraints following *step 2* and *step 3*:

1. On the **Flow Navigator** page, click on **Generate Block Design**. The following window will open. Use the default values, except for the number of jobs. These will make use of parallel processing using several processors, up to the ones dedicated to the Linux UbuntuVM, or the number of physical CPUs on your machine if you are using a native OS. Click **Generate**.

Figure 7.44 – Generating the ETS SoC project RTL file

2. To set the ETS SoC project synthesis options, including the synthesis constraints, in the Vivado GUI main menu, go to **Flow**, then **Settings**, and then select **Synthesis Settings…**, as indicated by the following figure.

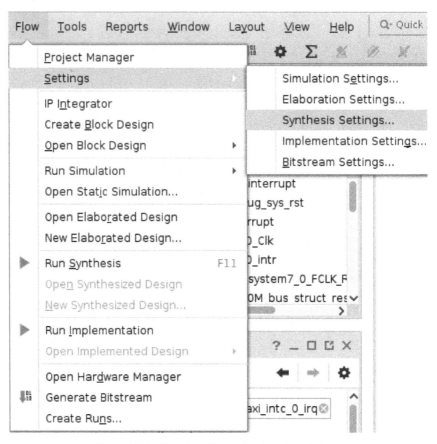

Figure 7.45 – Opening the synthesis entry window

3. The following **Synthesis** menu will open, where you can examine the different constraints available. When a constraint is selected, a description is provided at the bottom, as indicated in the following figure. Leave the options at their default values and click **OK**.

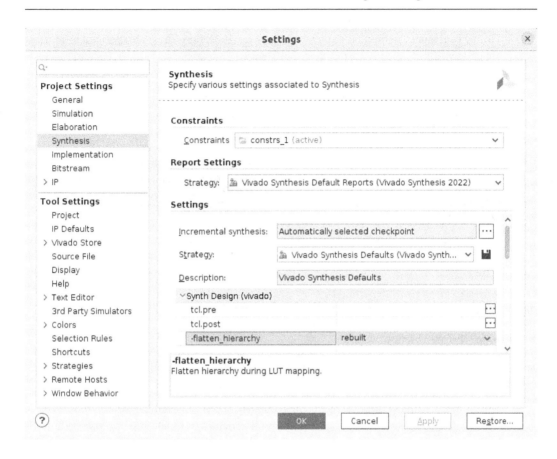

Figure 7.46 – Setting the synthesis constraints for the ETS SoC project

Implementation tool parameters affecting the PPA

Once the SoC design has been synthesized and the RTL has been converted into an FPGA-specific netlist, guided by the synthesis tools option to meet our PPA strategy, the next step is to implement the logical netlist to the physical elements of the FPGA resources. This physical implementation and mapping is a complex software-driven process with many optimization strategies, which are also guided by the user via design constraints. There are three types of FPGA implementation constraints:

- Physical constraints, defining the mapping between the logical elements in the netlist and their physical equivalent in the FPGA resources. These include the I/O pins, the physical placement of block RAMs, and the device configuration settings.

- Timing constraints, defining the clock frequency for the design. These are specified in a format defined by Xilinx called **Xilinx Design Constraints** (**XDCs**).

- Power constraints, defining the parameters for the power analysis. They include the operating conditions (voltage, power, and current settings) and the operating environment. They also specify the switching activity rates for the design's physical objects (nets, pins, block RAMs, transceivers, and DSP blocks).

These design constraints help to drive the Vivado implementation tools at every stage of the process and guide them toward an optimization strategy, in terms of logic, power, and physical placements.

The physical design within modern FPGA devices is becoming an engineering discipline worth the specialism, considering what it requires in terms of techniques and the learning curve. You are encouraged to explore this domain further using *Vivado Design Suite User Guide: Implementation for Xilinx*, available at `https://docs.xilinx.com/viewer/book-attachment/ FwU2AZhBjPWyTDgTpfvMrg/3OSTgfIjc~pSJ2lU_YabuQ`.

Specifying the implementation options for the ETS SoC design

The steps for this are as follows:

1. To enter the implementation options, like for the synthesis settings specification, go to **Flow**, then **Settings**, and then select **Implementation Settings**.

2. The following **Implementation** menu will open, where you can examine the different available settings. When a row is selected, a description is provided at the bottom, as indicated in the following figure. Leave the options at their default values and click **OK**.

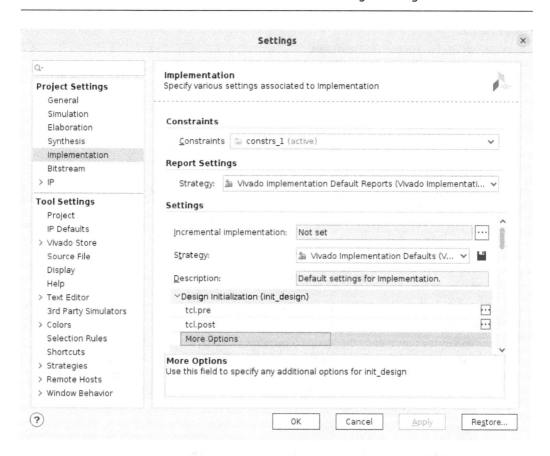

Figure 7.47 – Setting the implementation settings for the ETS SoC project

Specifying the implementation constraints for the ETS SoC design

The steps for this are as follows:

1. To enter the implementation constraints, in the Vivado GUI Flow Navigator, go to **Implementation | Open Implemented Design | Constraints Wizard**. We already have the timing constraints defined for the preset template, which we kept for the ETS SoC design and can be examined in *Step 2*.

2. To enter or view the timing constraints, in the Vivado GUI Flow Navigator, go to **Implementation | Open Implemented Design | Edit Timing Constraints**. This will open the following window, where the timing constraints can be edited.

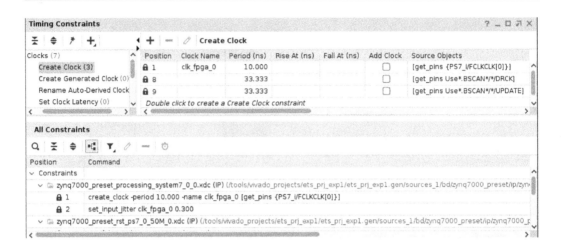

Figure 7.48 – Editing the timing constraints for the ETS SoC project

> **Information**
>
> You can specify the timing and other design constraints for both the synthesis and implementation and add them as an XDC file to the design project. For more information, please check out *Vivado Design Suite User Guide: Using Constraints*, available at https://docs.xilinx.com/viewer/book-attachment/4dVZhvUrG0H01LbNfj76YA/VjEOZSdbHCzk6bRsAvMLfg.

SoC hardware subsystem integration into the FPGA top-level design

Since we have started the ETS SoC design from a template project that already had a top-level HDL wrapper generated (zynq7000_preset_wrapper.v), the project is already set for a full implementation flow, and there is no need to instantiate it within a higher-level design. Conversely, if the project was started from scratch and we used IP Integrator to generate the design, as we did in *Chapter 2, FPGA Devices and SoC Design Tools*, we could simply right-click on the IP-generated project and select **Create HDL Wrapper** to generate the top-level RTL, or use it as an instantiation template at a higher level of hierarchy. The following figure illustrates how to create the *HDL wrapper* for the IP design.

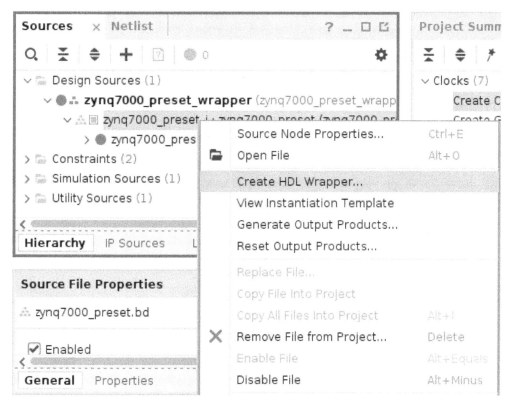

Figure 7.49 – Generating an HDL wrapper for instantiation in a higher level of hierarchy project

Verifying the FPGA SoC design using RTL simulation

The ETS SoC design project was started from a template design created by Xilinx Vivado for a known board, and then it was customized to add the PP into the PL block. The preset design already had a test bench included to test the AXI GPIO and AXI BRAM that initially formed the PL block. We can keep the same test bench for our simulation purposes, but since we have customized the address map of the ETS SoC design, we need to adjust the addresses used in the test bench to their new values. We can also extend it to verify other PP IPs to ensure their correct hardware functionality and integration. The test bench uses an AXI **Verification IP (VIP)**, which is provided by Xilinx in the Vivado verification library to test the proper functioning of connectivity between AXI masters and slaves in the custom RTL design flow, such as in the ETS SoC project. We can add a test for the AXI INTC IP, which adds the IPC interrupts from the MicroBlaze to the Cortex-A9 processor.

> **Information**
>
> More information on the Xilinx AXI VIP is available at `https://www.xilinx.com/products/intellectual-property/axi-vip.html`.

Customizing the ETS SoC design verification test bench

Let's customize the test bench provided in the ETS SoC initial design template by following the following steps:

1. In the ETS SoC project within the Vivado GUI, locate the test bench source code, as illustrated here, and then double-click it to open it in the Vivado Source Code Editor.

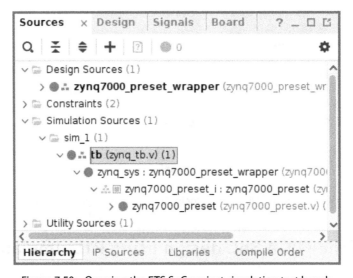

Figure 7.50 – Opening the ETS SoC project simulation test bench

2. The AXI GPIO IP in the ETS SoC project address map is still at **0x4120_0000**; however, we have moved the AXI BRAM from **0x4000_0000** to **0x4200_0000**. In (`zynq_tb.v`), edit the lines accessing the AXI BRAM and change the address used for both the write and read statements from **32'h40000000** to `32'h42000000`.

3. We can also test the IPC interrupt mechanism between the MicroBlaze and the Cortex-A9 processor using the AXI INTC SW-generated interrupt mechanism. To generate an SW interrupt using the AXI INTC, we first need to enable (unmask) the SW interrupt using the IER register. The IER register is at the AXI INTC, `[[base address] + 08h]`. The IER register has eight bits we selected for the SW-generated interrupts at its customization stage. It has no hardware pins or interrupt-connected inputs. Now, to generate an SW interrupt using the AXI INTC, we simply need to write to the corresponding bit of the ISR register, which is one of the ISR[7:0] bits. The ISR register is at offset 0x00 from the AXI INTC base address. The base address of the

AXI INTC is at **0x4240_0000**. In the test bench (`zynq_tb.v`), add the following lines after the end of the AXI GPIO test and just before the start of the AXI BRAM test:

```
// Testing the AXI INTC Interrupt Controller
$display ("Writing to the AXI INTC IER[0] to Enable the
SW IRQ");
tb.zynq_sys.zynq7000_preset_i.processing_system7_0.inst.
write_data( 32'h42400008,4, 32'h1, resp);
repeat(10) @(posedge tb_ACLK);
$display ("Writing to the AXI INTC to generate SW IRQ");
tb.zynq_sys.zynq7000_preset_i.processing_system7_0.inst.
write_data( 32'h42400000,4, 32'h1, resp);
repeat(10) @(posedge tb_ACLK);
$display ("Reading from the AXI INTC to check that
the SW triggered Interrupt is pending"); tb.zynq_sys.
zynq7000_preset_i.processing_system7_0.inst.read_
data(32'h42400004,4,read_data,resp);
repeat(10) @(posedge tb_ACLK);
if(read_data == 32'h1) begin
    $display ("AXI VIP SW Interrupt Generation Test
PASSED");
end
else begin
    $display ("AXI VIP SW Interrupt Generation Test
FAILED");
end
```

4. Let's go back to the IP Integrator window and make sure that the PS-PL configuration matches the following figure to avoid an unresolved issue in the AXI VIP.

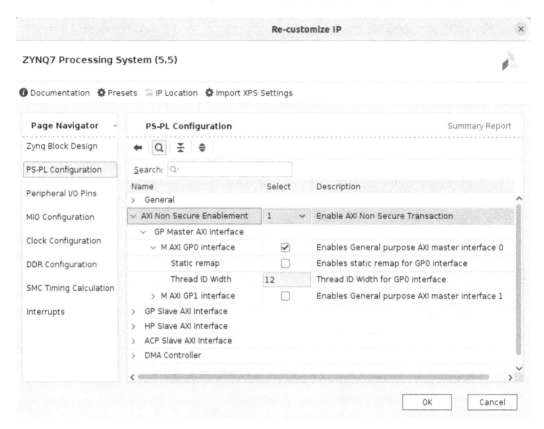

Figure 7.51 – The PS-PL GP slave AXI interface settings to use for the ETS SoC design

> **Information**
>
> There is a known simulation issue with the AXI VIP that we can hit if we don't set the PS-PL GP slave AXI interface as shown in the previous figure. For further details on this issue, please check out https://support.xilinx.com/s/question/0D52E00006hp1h2SAA/ axi-simulation-bug?language=en_US.

Hardware verification of the ETS SoC design using the test bench

We have thus far explored the test bench provided by the Vivado design template, then customized it to match our settings, and finally, expanded it to use the AXI VIP to test the IPC interrupt mechanism. We have chosen a microarchitecture implementation that uses a doorbell register to generate the IPC interrupt, from the MicroBlaze in the PP within the PL block to the Cortex-A9 within the PS block. We can now load the test bench and run the simulation in Vivado:

1. In the Vivado Flow Navigator, select **Simulation | Run Simulation | Run Behavioral Simulation**.

2. The test bench will be loaded, and the simulation will start up to the initialization stage. To run the simulation, in the Vivado GUI main menu, go to **Run | Run All**. This will run the simulation to completion. We have two sets of results to examine, the simulation waveform and the Tcl Console printing the output of the display statements from the test bench.

3. The simulation waveform looks like the following figure. We can see the transactions from the AXI GPIO, then the AXI INTC, and then the last one from the AXI BRAM.

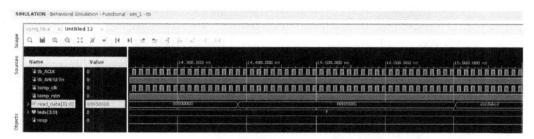

Figure 7.52 – The ETS SoC design RTL simulation waveform

4. The simulation console output is more verbose and displays the transaction flow and the display statement from the test bench. The tests are passing, and the results are as expected. It looks like the following:

```
Tcl Console   ×  Messages  | Log                                                               ?  _ □ ⊡

Q  ≍  ♦  II  |  ⊟  |  ⊞  |  🗑

XilinxAXIVIP: Found at Path: tb.zynq_sys.zynq7000_preset_i.processing_system7_0.inst.S_AXI_ACP.slave
[0] : *ZYNQ_VIP_INFO : S_AXI_ACP : Port is DISABLED.
running the tb
                0 else checking line ......0
[810] : *ZYNQ_VIP_INFO : FPGA Soft Reset called for 0x1
[910] : *ZYNQ_VIP_INFO : FPGA Soft Reset called for 0x0
INFO: [USF-XSim-96] XSim completed. Design snapshot 'tb_behav' loaded.
INFO: [USF-XSim-97] XSim simulation ran for 1000ns
launch_simulation: Time (s): cpu = 00:03:19 ; elapsed = 00:01:13 ; Memory (MB): peak = 9638.184 ; gain = 0.000 ; free physical = 1117 ; free virtual = 2909
run all
[12910] : M_AXI_GPO : *ZYNQ_VIP_INFO : Starting Address(0x41200000) -> AXI Write -> 4 bytes
wr_id called with wr_size 4b7
            12910ID1 in strb task is 991
[13095] : M_AXI_GPO : *ZYNQ_VIP_INFO : Done AXI Write for Starting Address(0x41200000) with Response 'OKAY'
LEDs are toggled, observe the waveform
Writing to the AXI INTC IER[0] to Enable/Unmask the SW triggered Interrupt
[13295] : M_AXI_GPO : *ZYNQ_VIP_INFO : Starting Address(0x42400008) -> AXI Write -> 4 bytes
wr_id called with wr_size d3d
            13295ID1 in strb task is 583
[13475] : M_AXI_GPO : *ZYNQ_VIP_INFO : Done AXI Write for Starting Address(0x42400008) with Response 'OKAY'
Writing to the AXI INTC to generate SW triggered Interrupt to the Cortex-A55
[13670] : M_AXI_GPO : *ZYNQ_VIP_INFO : Starting Address(0x42400000) -> AXI Write -> 4 bytes
wr_id called with wr_size 7ca
            13670ID1 in strb task is 841
[14005] : M_AXI_GPO : *ZYNQ_VIP_INFO : Done AXI Write for Starting Address(0x42400000) with Response 'OKAY'
Reading from the AXI INTC to check that the SW triggered Interrupt is pending
[14190] : M_AXI_GPO : *ZYNQ_VIP_INFO : Starting Address(0x42400004) -> AXI Read -> 4 bytes
            14190ID2 in read strb task is da2
[14375] : M_AXI_GPO : *ZYNQ_VIP_INFO : Done AXI Read for Starting Address(0x42400004) with Response 'OKAY'
AXI VIP SW Interrupt Generation for the Cortex-A9 Test PASSED
[14570] : M_AXI_GPO : *ZYNQ_VIP_INFO : Starting Address(0x42000000) -> AXI Write -> 4 bytes
wr_id called with wr_size 743
            14570ID1 in strb task is 8ad
[14675] : M_AXI_GPO : *ZYNQ_VIP_INFO : Done AXI Write for Starting Address(0x42000000) with Response 'OKAY'
[14875] : M_AXI_GPO : *ZYNQ_VIP_INFO : Starting Address(0x42000000) -> AXI Read -> 4 bytes
            14875ID2 in read strb task is 382
[15005] : M_AXI_GPO : *ZYNQ_VIP_INFO : Done AXI Read for Starting Address(0x42000000) with Response 'OKAY'
15205 ns, running the testbench, data read from BRAM was 32'habcdabcd
AXI VIP Test PASSED
Simulation completed
$stop called at time : 15205 ns : File "/tools/vivado_projects/ets_prj_expl/ets_prj_expl.srcs/sim_1/imports/zynq7000_preset/zynq_tb.v" Line 103
<                                                                                                >

Type a Tcl command here
```

Figure 7.53 – The ETS SoC design RTL simulation Tcl Console output

Implementing the FPGA SoC design and FPGA hardware image generation

As already introduced in *Chapter 2, FPGA Devices and SoC Design Tools,* of this book, the design implementation takes the design netlist and the user constraints (including the implementation settings) as input and produces the physical design netlist. This physical netlist is then mapped, placed, and routed using the FPGA hardware resources and meeting (when possible) the user constraints. Once the physical netlist is produced, the FPGA image or the bitstream file is generated to configure the FPGA device. The FPGA device PL block can be programmed via JTAG when still in the debugging stages of the design, or by the PS. In the debugging stages of the SoC, the configuration is done through the Xilinx FPGA JTAG interface. The bitstream is then downloaded directly to the FPGA PL block from the host development machine using a JTAG cable, connecting the host machine to the FPGA board.

ETS SoC design implementation

We have already introduced the implementation constraints and settings in the PPA section of this chapter. The only action we need to perform now to implement the design, once its RTL files have been synthesized and its netlist has been produced, is to launch the implementation step from the Vivado Flow Navigator by selecting **Implementation | Run Implementation**. This will take some time to accomplish, according to how much system memory and how many CPUs are used to perform the task, as the jobs associated with the implementation can be parallelized. If the design can meet the specified constraints, then the implementation will finish with no errors, and we can then proceed to generate the FPGA bitstream.

ETS SoC design FPGA bitstream generation

To open the FPGA bitstream file generation options menu, in the Vivado GUI, go to **PROJECT MANAGER** in the **Flow Navigator** section, and then select **Settings**. Once the **Settings** menu is open, select the **Bitstream** row, which will open the following window.

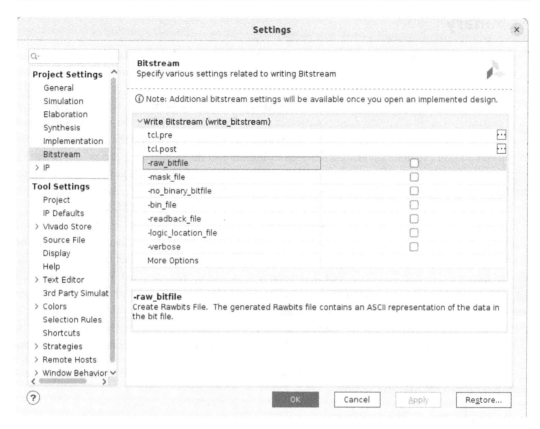

Figure 7.54 – The FPGA bitstream file generation options

In the preceding menu, most of the options are related to the FPGA configuration operational control, such as during a cyclic readback of the FPGA bitstream once deployed in the field. The read-back file from the FPGA is then compared to a known good bitstream file stored on non-volatile memory to make sure that it didn't get corrupted due to environmental issues, such as **Single-Event Upsets** (**SEUs**).

Once both the hardware and the software designs are debugged and have reached a certain level of development maturity, the FPGA bitstream can be stored in non-volatile memory on the electronics board, from which the FPGA is automatically configured on the system power-up by the PS. We will cover the configuration and boot modes later in this book.

> **Information**
>
> The following article provides details on the supported non-volatile storage for booting and configuring the Zynq-7000 SoC FPGAs: `https://support.xilinx.com/s/article/50991?language=en_US`

Summary

In this chapter, we started by providing some operational and practical guidance on how to install and use the Xilinx SoC development tools on an Ubuntu Linux VM, built and hosted on Oracle VirtualBox. We revisited the ETS SoC architecture requirements and defined a microarchitecture based on a PP engine that uses a MicroBlaze processor subsystem to implement the Ethernet frames filtering hardware acceleration. We used a Vivado example project preset for a known Xilinx Zynq-7000 SoC board as a starting template design, and we also customized the PS block to match the microarchitecture requirements. We extended the template design by building all the required functionalities in the PL block, using IPs from the Vivado library. We went through a full customization process to design the hardware of an embedded system in the PL block based on the MicroBlaze. Additionally, we put in place the necessary infrastructure for the IPC between the Cortex-A9 and MicroBlaze as needed by the envisaged ETS SoC software architecture, which we will implement in *Chapter 8, FPGA SoC Software Design Flow*. We also covered hardware verification using an RTL test bench to simulate the ETS SoC design, and we checked that the introduced customization didn't break the initial template design functionality and that the added IPC features were working as expected. We also introduced the PPA concept and its important aspects for a successful SoC project. We looked at the FPGA design constraints for both the synthesis and the implementation stages, how to specify the settings for these flows, and how to enter the design constraints that influence the achieved implementation results. Finally, we covered the design implementation flow, from synthesis to the FPGA bitstream generation.

The next chapter will continue by using the same practical approach for the software implementation flow for the ETS SoC design.

Questions

Answer the following questions to test your knowledge of this chapter:

1. How is communication established between the main ETS SoC software and the hardware accelerator? Are there any alternative approaches you can think of?

2. How can we augment the capabilities of the proposed microarchitecture and scale it for future needs?

3. Why is the TS field used in the ETMP UDP packet?

4. Which field in the ETMP UDP packet is processed better in hardware instead of the MicroBlaze software? Why?

5. What are the advantages of starting the ETS SoC design from a template preset?

6. Describe the steps needed to augment the number of IPC interrupts between the Cortex-A9 and the MicroBlaze processors from 8 to 16 interrupts.

7. What is the frequency we chose to run the PL logic at? How can we increase it to 125 MHz?

8. How can we check that the aforementioned increase of the PL logic frequency to 125 MHz is okay for the ETS SoC project?

9. How are the PL interrupts targeting the Cortex-A9 connected?

10. How can we test that the design customization performed in IP Integrator is in line with the Vivado tools' expectations for the ETS SoC project?

11. In the address map view of IP Integrator, why do we have two networks?

12. In the address map, under Network_0, why do we have two address maps?

13. How can PPA influence the success (or lack of suuccess) of the ETS SoC project?

14. What are the design constraints? Why do we use them? How can we specify them in Vivado for the ETS SoC project?

15. What were the changes needed to customize the preset RTL test bench for the ETS SoC project?

16. How can we check that the design is meeting the expected functionality using simulation?

17. What is an FPGA bitstream? And how can it be loaded to the FPGA device?

18. How could an FPGA-based SoC design get corrupted due to its bitstream while deployed in the field? How could we monitor and correct such behavior?

8

FPGA SoC Software Design Flow

In this chapter, we will delve into the implementation phase of the SoC software of the **Electronic Trading System** (**ETS**) for which we developed the architecture in *Chapter 6, What Goes Where in a High-Speed SoC Design*, and built the hardware in *Chapter 7, FPGA SoC Hardware Design and Verification Flow, FPGA SoC Hardware Design and Verification Flow*. We will define the SoC software microarchitecture for both the Cortex-A9 processor and its accelerator, the MicroBlaze **Packet Processor** (**PP**). We will explore the embedded software development flow using the Xilinx Vitis environment and how to write simple software to run on the SoC processors. We will mainly use the Vitis IDE-generated test application source code for the peripherals included in the design to understand how to configure, access, and then use them. This exercise will prepare you to write more complex software applications for the ETS SoC design in *Part 3*. This chapter is mainly hands-on and you will be guided at every step of the SoC software design phases from the concept to executable image generation using the Vitis IDE.

In this chapter, we're going to cover the following topics:

- Major steps of the SoC software design flow

- Setting up the BSP, boot software, drivers, and libraries for the software project

- Defining the distributed software microarchitecture for the ETS SoC processors

- Building the user software applications to initialize and test the SoC hardware

Technical requirements

The GitHub repo for this title can be found here: `https://github.com/PacktPublishing/Architecting-and-Building-High-Speed-SoCs`.

Code in Action videos for this chapter: `http://bit.ly/3hfoir2`.

Major steps of the SoC software design flow

As previously introduced in *Chapter 2, FPGA Devices and SoC Design Tools*, the software development for the Xilinx FPGA SoC is performed using the Vitis tools. A project for the ETS SoC is first created in the Vitis IDE using its XSA archive file – this file needs to be generated by the Vivado IDE for the ETS SoC hardware.

The full flow of the software design process in the Vitis IDE is summarized by the following diagram:

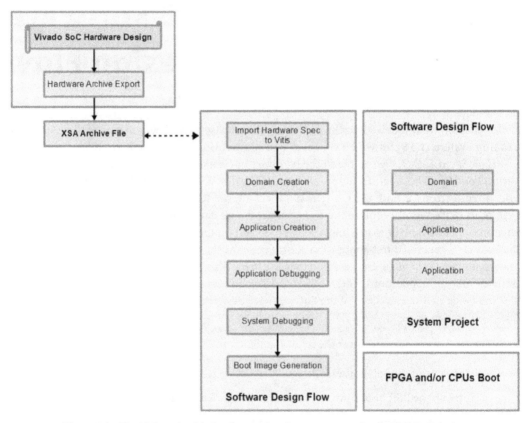

Figure 8.1 – The Vitis embedded software development steps for the ETS SoC design

ETS SoC XSA archive file generation in the Vivado IDE

First, we need to generate the XSA file within the Vivado IDE by following these steps:

1.   Open the ETS SoC design in Vivado and then go to **File | Export | Export Hardware Platform** as shown by the following figure:

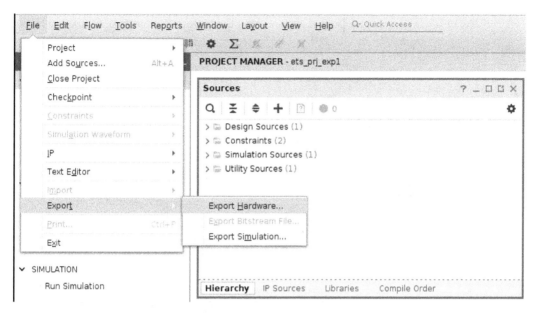

Figure 8.2 – Accessing the Vivado XSA file generation wizard

2.   The **Export Hardware Platform** wizard will open. Set **Platform Type** to **Fixed**, and then click **Next**:

Figure 8.3 – The Vivado XSA file generation welcome screen

3. In the next window, for the **Output** option, select the **Include bitstream** option so the FPGA can be programmed from Vitis IDE, as illustrated by the following figure. Click **Next**:

Figure 8.4 – Vivado XSA file generation options

4. This **Files** window is used to set the name and the location of the XSA file for the ETS SoC design. Set the desired values as shown and then click **Next**:

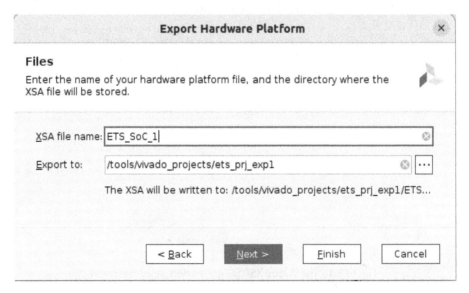

Figure 8.5 – The Vivado XSA file specification

5.   The following window provides a summary of the chosen settings for the XSA file generation – review these values and then click **Finish** to generate the XSA file:

Figure 8.6 – A summary of the Vivado XSA file generation

ETS SoC software project setup in Vitis IDE

Once the XSA archive file has been created for the ETS SoC hardware design, we can use the Vitis IDE to import the ETS SoC hardware specification into the Vitis environment, which will allow us to work on the software development part of the ETS SoC. Let's begin:

1.   Launch the Vitis IDE using the following command line:

```
$ cd <Tools_Install_Directory>/Xilinx/Vivado/2022.1/bin/
$ ./vivado
```

Replace <Tools_Install_Directory> with the path where you have installed Vitis on your machine or the UbuntuVM Linux VM if you are using it as a host.

2.   Once Vitis is up and running, we can then create the ETS SoC Vitis project using the XSA archive file of the hardware design we have produced in the Vivado IDE. When Vitis is launched, first specify a workspace for the Vitis environment, as shown in the following diagram, and then click **Launch**:

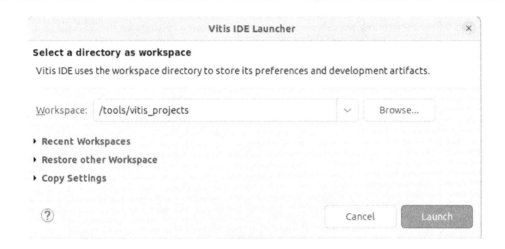

Figure 8.7 – Launching Vitis IDE and specifying its workspace directory

3. From the Vitis IDE **Welcome screen** and under **Project**, click on **Create Application Project** as shown:

Figure 8.8 – Launching the Create Application Project menu in the Vitis IDE

4. An introduction screen for the Vitis IDE **Create Application Project** wizard opens as shown. Review the content to refresh the information about the Vitis IDE project structure already introduced in *Chapter 2, FPGA Devices and SoC Design Tools*. Once done, click **Next**:

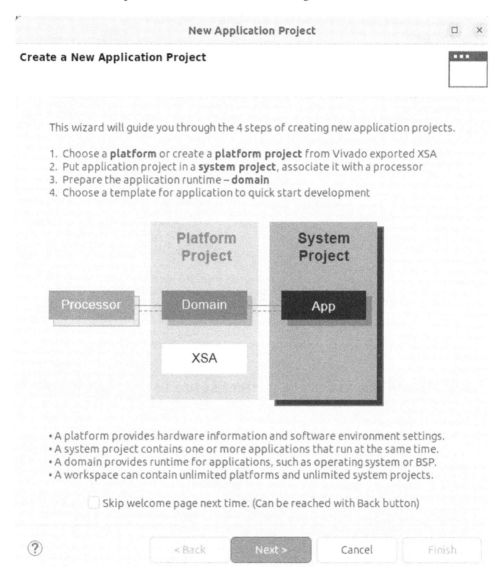

Figure 8.9 – The Vitis IDE project structure information

5. Select the **Create new platform from hardware (XSA)** tab, specify the ETS SoC XSA archive file location, and fill in the **Platform name** field as shown. Click **Next**:

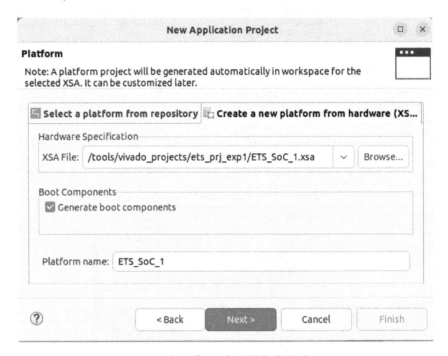

Figure 8.10 – Specifying the ETS SoC XSA location

ETS SoC MicroBlaze software project setup in the Vitis IDE

Once the new platform has been created in the Vitis IDE using the XSA hardware archive file we imported from Vivado, we can start the process of creating the software projects and their corresponding domains. We can start with any processor detected in the hardware platform by the Vitis IDE. Let's start with the MicroBlaze PP of our ETS SoC project:

1. Select the MicroBlaze processor hardware name instance as highlighted in the following figure. Provide the details by specifying the **Application project name**, **System project name**, and **Target processor** information to associate with the project as shown. Then, click **Next**.

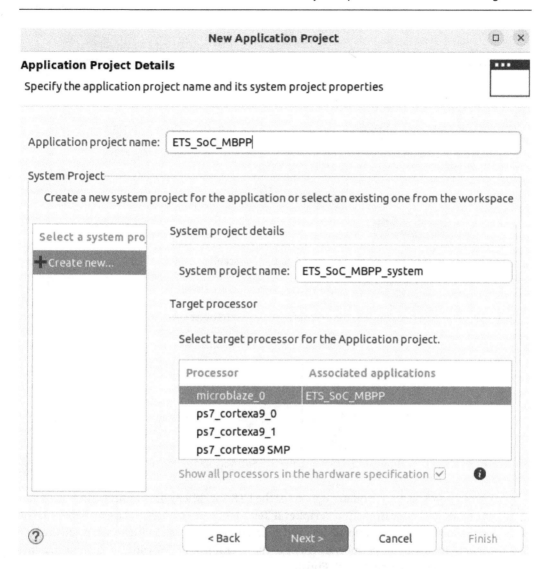

Figure 8.11 – Specifying the ETS SoC MicroBlaze application project details

2. Now, we can create the domain to which the ETS SoC MicroBlaze application project will link as shown. Click **Next**:

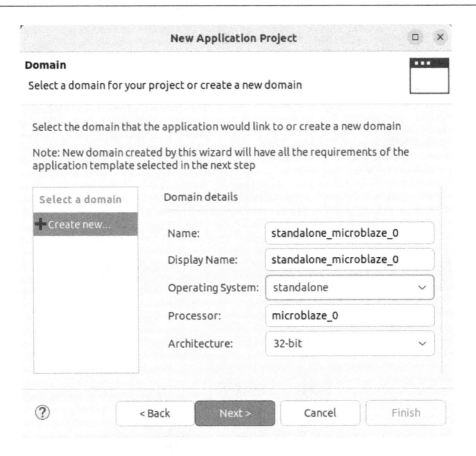

Figure 8.12 – Creating the ETS SoC MicroBlaze domain

3.   We can now select a **Templates** option for the ETS SoC MicroBlaze project. There are many examples to choose from. The **Peripheral Tests** template is a useful starting point – it will give us all the necessary information and code snippets that we can use to set, configure, and communicate with the peripherals visible to the MicroBlaze PP. We can also examine that they are operating as expected. Click **Finish**:

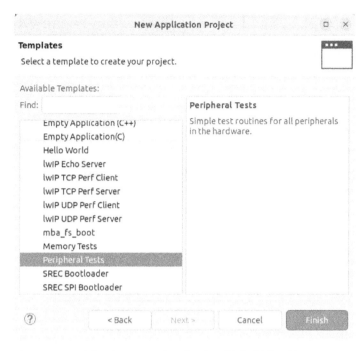

Figure 8.13 – Selecting a template for the ETS SoC MicroBlaze project

4. The domain and its associated project are now created and visible in the Vitis IDE as shown. You can examine their structure and content to gain some initial familiarity with them:

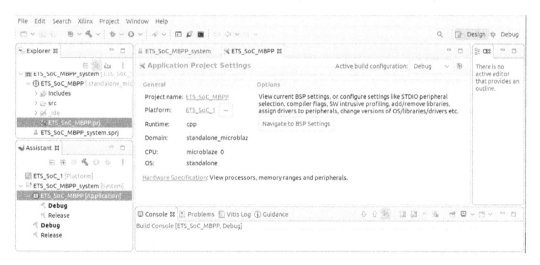

Figure 8.14 – An overview of the MicroBlaze ETS SoC project in Vitis

ETS SoC PS Cortex-A9 software project setup in the Vitis IDE

To create a second project for the ETS SoC Cortex-A9 processor in Vitis IDE, we need to create a second domain to which this second project will be linked first – then, we create the application project for the Cortex-A9 following almost the same steps as we did for the MicroBlaze PP. The only difference is that we don't have to specify a new platform in Vitis, as it is already created:

1. First, double-click on `platform.spr` as shown:

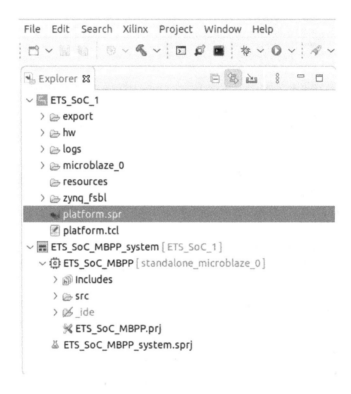

Figure 8.15 – Opening the ETS_SoC_1 platform in the Vitis IDE

2. This will open the platform summary page in the Vitis IDE. Click the + sign to start a new domain creation linked to this platform. The **New Domain creation** wizard will be launched. Specify the information as entered in the following screenshot and click **OK**:

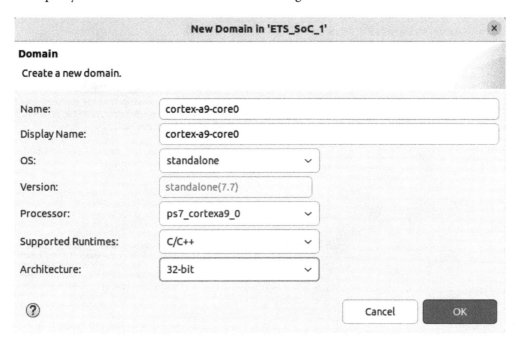

Figure 8.16 – Creating a new domain for the ETS SoC in the Vitis IDE

3. The **New Application Project** wizard will open. Choose the first tab, **Select a platform from the repository**, as we already have created the ETS SoC platform using the XSA hardware archive file in previous steps. Select **ETS_SoC_1 [custom]** as shown and click **Next**:

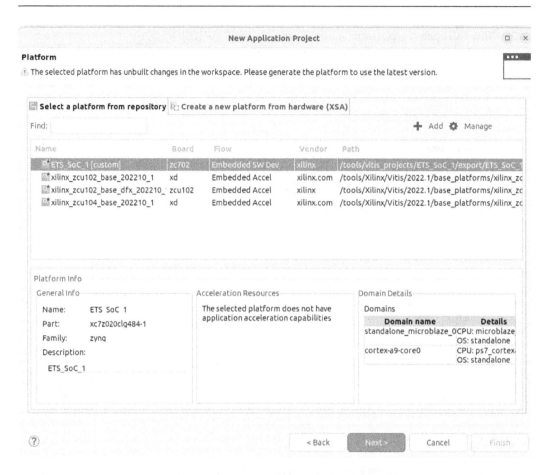

Figure 8.17 – Selecting the ETS_SoC_1 platform in the Vitis IDE

4. As with when we created the new MicroBlaze PP application project, specify the details, now for the Cortex-A9 core0 project instead, as shown, and click **Next**:

Major steps of the SoC software design flow   249

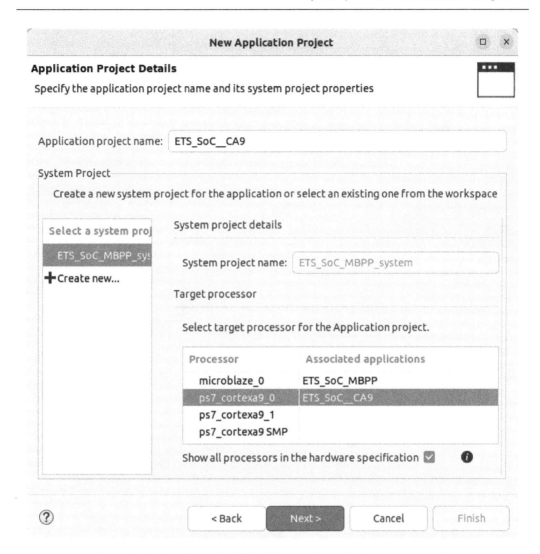

Figure 8.18 – Specifying the ETS SoC Cortex-A9 application project details

5. Now, we can select the domain to which the ETS SoC Cortex-A9 **New Application Project** will be linked. We have just created this domain, `cortex-a9-core0`, in *step 2*. Click **Next**:

Figure 8.19 – Selecting the ETS SoC Cortex-A9 domain

6. We can now select a template for the ETS SoC Cortex-A9 core0 software project. The **Peripheral Tests** template is a useful project to get familiarity with the Xilinx device drivers. Select the **Peripheral Tests** template for now and click **Finish**:

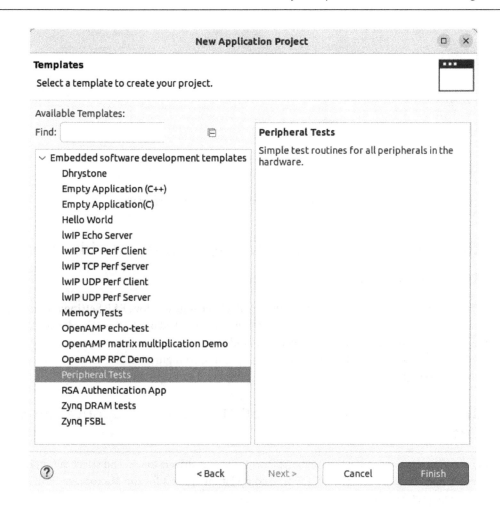

Figure 8.20 – Selecting a template for the ETS SoC Cortex-A9 project

7. The second domain and its associated software project are now created for the Cortex-A9 core0 processor and visible in the Vitis IDE. You can also examine their structure and content to gain some initial familiarity with them:

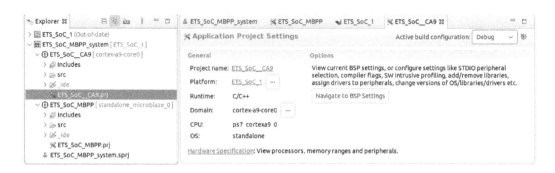

Figure 8.21 – An overview of the ETS SoC projects in Vitis

Setting up the BSP, boot software, drivers, and libraries for the software project

As can be seen in *Figure 8.21*, in the **Application Project Settings** window, **BSP Settings** is accessible from the Vitis IDE per application project. Also, when we first specified our ETS SoC hardware platform, by using the XSA hardware archive generated by Vivado, we selected **Generate boot components** (as in *Figure 8.10*). We should easily accomplish the remaining configuration and settings tasks for the boot, the **Board Support Package** (**BSP**), and the peripheral software drivers.

Setting up the BSP for the ETS SoC MicroBlaze PP application project

Within the Vitis IDE, we can customize the BSP, set the device drivers to use, and select the application libraries we need. We can also specify the BSP compilation options for the MicroBlaze PP ETS SoC application project. Let's go through it by following these steps:

1.   To access the MicroBlaze PP application project in the Vitis IDE, simply expand the **ETS_SoC_1** platform and select **Board Support Package** under the **microblaze_0** entry, as shown in the following figure:

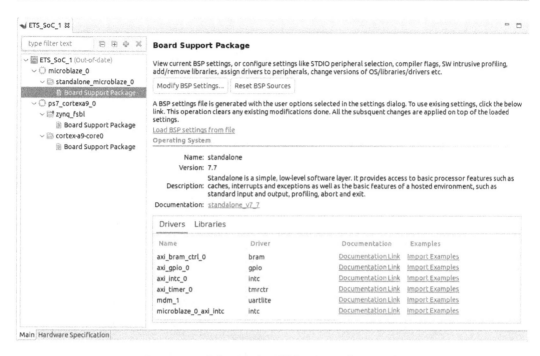

Figure 8.22 – Selecting the ETS SoC MicroBlaze PP BSP

2. To customize the BSP for the MicroBlaze PP application project, click on **Modify BSP Settings…** – this will open the following window:

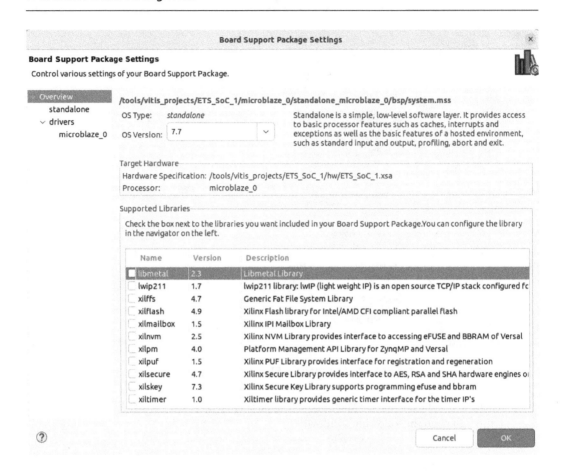

Figure 8.23 – Selecting the libraries for the MicroBlaze PP application project

3. Select the standalone row to open the **Board Support Package Settings** window where we can specify the **stdin** and **stdout** devices, and all the configurations related to the **Operating System (OS)**, which is *baremetal* in our case:

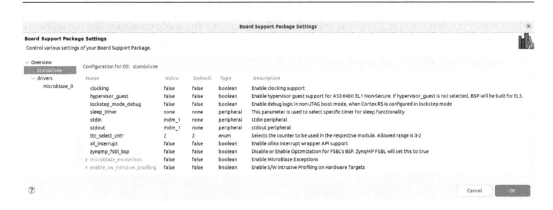

Figure 8.24 – Specifying the BSP settings for the MicroBlaze PP OS

4. Select the drivers row to open the **Drivers** settings window as shown. Make sure that every device has an associated driver selected, as set by default:

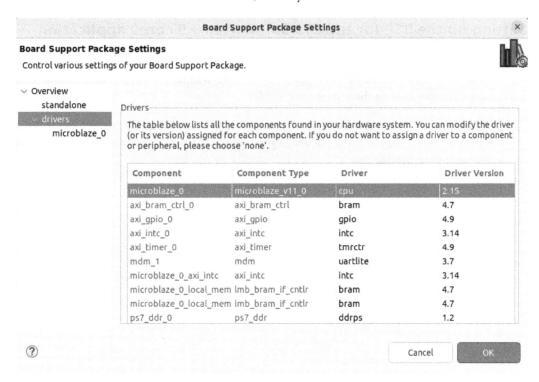

Figure 8.25 – Specifying the device drivers for the MicroBlaze PP application project

5. Select the **microblaze_0** row to open the BSP settings for the tools used to build the software and their options. Leave the default values as set by the Vitis IDE and click **OK**. This will create the necessary BSP package, as selected in the preceding steps:

Figure 8.26 – Specifying the build tools options for the MicroBlaze PP application project

Setting up the BSP for the ETS SoC Cortex-A9 core0 application project

The steps are the same as for the MicroBlaze PP, although the settings and options are different. Let's go through them by following these steps:

1. To access the **Cortex-A9 core0** application project in the Vitis IDE, simply expand the **ETS_SoC_1** platform and select **Board Support Package** under the **cortex-a9-core0** entry as shown in the following figure:

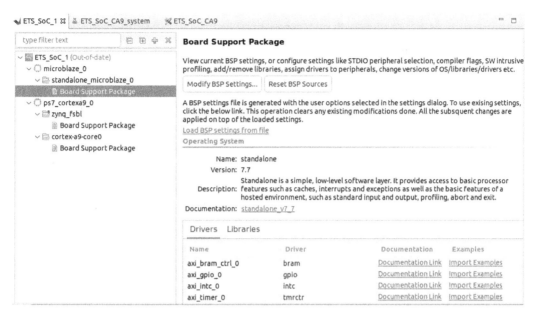

Figure 8.27 – Selecting the ETS SoC Cortex-A9 core0 BSP

2. To customize the BSP for the **Cortex-A9 core0** application project, click **Modify BSP Settings...**
    – this will open the following window. Select the lwIP211 library – we may choose to use its
    services to implement the UDP client for the Cortex-A9 processor for communication functions
    with the **Electronic Trading Market** (**ETM**) over the Ethernet:

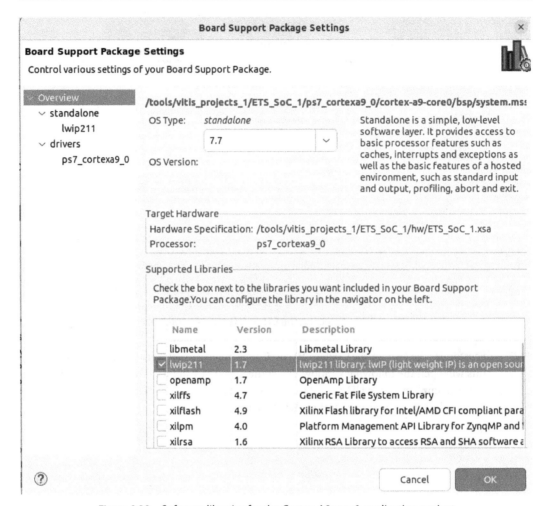

Figure 8.28 – Software libraries for the Cortex-A9 core0 application project

3. Select the standalone row to open the **Board Support Package Settings** window where we can specify the **stdin** and **stdout** devices, and all the configurations related to the OS, which is *baremetal* (or a standalone one) in our case:

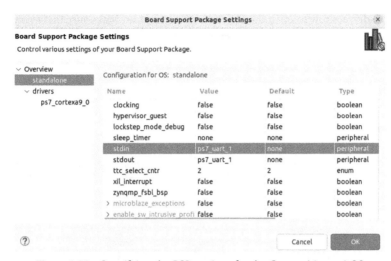

Figure 8.29 – Specifying the BSP settings for the Cortex-A9 core0 OS

4.   Through this window, we can select any software library we need for the application software. Leave all the settings as their default values:

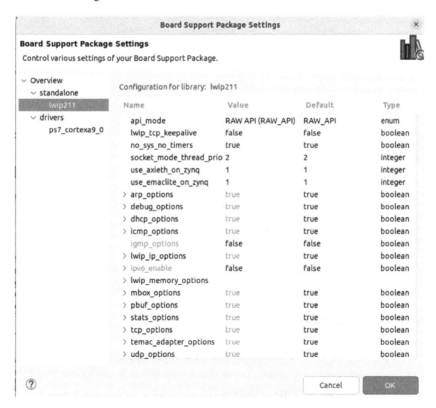

Figure 8.30 – Specifying the lwIP TCP/IP stack for the Cortex-A9 core0 OS

5. Select the **drivers** row to open the **Drivers** setting window. Make sure that every device has an associated driver with it selected, as set by default:

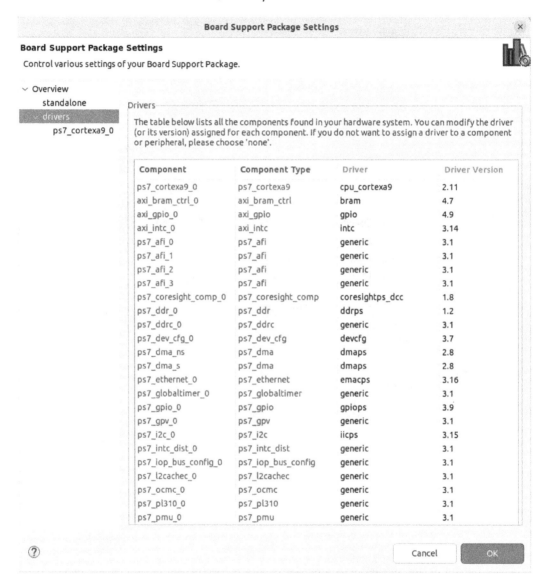

Figure 8.31 – Specifying the device drivers for the Cortex-A9 core0 application project

6. Select the **ps7_cortex_a9_0** row to open **Board Package Support Settings** for the tools used to build the software and their options. Leave the default values as set by the Vitis IDE and click **OK**. This will create the necessary selected BSP package:

Figure 8.32 – Specifying the build tool options for the Cortex-A9 core0 application project

Setting up the BSP for the ETS SoC boot application project

When we first specified our ETS SoC hardware platform, by using the XSA hardware archive generated in the Vivado IDE, we selected **Generate boot components** as shown in *Figure 8.10*. As you may have noticed, this has automatically created an application project associated with the Cortex-A9 core0 and for which a BSP is also provided. We will just examine its content, so we know what is used to build such an application project to boot the system on powering up:

1. To access the boot project associated with the Cortex-A9 core0 in the Vitis IDE, simply expand the **ETS_SoC_1** platform and select **Board Support Package** under the **zynq_fsbl** entry, as shown by the following figure:

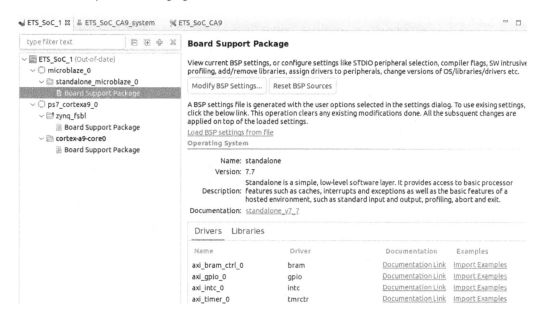

Figure 8.33 – Selecting the ETS SoC boot application BSP

2. This will open the BSP entry of the boot application project associated with the Cortex-A9 core0. We can see that the boot library settings use a *Generic Fat Filesystem*, as well as some *security software libraries*, provided by Xilinx and automatically set by the Vitis IDE. Leave the settings as their default values:

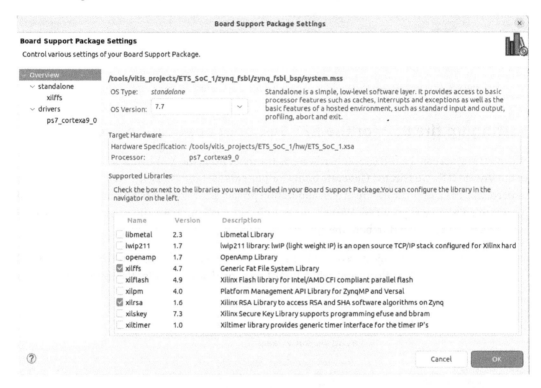

Figure 8.34 – Library settings for the Cortex-A9 core0 boot application project

3. Select the **ps7_cortex_a9_0** row to open the BSP settings for the tools used to build the software and their options as shown in the following figure. Leave the default values as set by the Vitis IDE and click **OK**. This will create the necessary BSP package, as selected in the preceding steps:

Figure 8.35 – Specifying the build tool options for the boot application project

Defining the distributed software microarchitecture for the ETS SoC processors

Thus far in this chapter, we have learned how a software project is created using the Vitis IDE, associated with a specific processor in the ETS SoC project, and how its BSP is configured. We can now delve into the software application-building process. We will develop a software microarchitecture for each processor core used in the ETS SoC design first. This will be based on the system architecture we developed in *Chapter 6, What Goes Where in a High-Speed SoC Design,* and the hardware implementation choices we made in *Chapter 7, FPGA SoC Hardware Design and Verification Flow,* such as the IPC mechanisms in both directions between the Cortex-A9 and the MicroBlaze PP processors. We can now revisit some remaining open items in the SoC system architecture. We have also defined the **Electronic Trading Market Protocol (ETMP)**; therefore, the filtering tasks are easily identifiable by reading the UDP packet payload of the ETMP. Let's start by redrawing the system hardware microarchitecture in a simplified view with the hardware implementation options we have made. We will also revisit the software-to-hardware communication model we created in *Chapter 6, What Goes Where in a High-Speed SoC Design,* and fill in any missing microarchitectural detail necessary for a correct and complete exchange of information between them.

A simplified view of the ETS SoC hardware microarchitecture

Following the ETS SoC initial system architecture definition, we have made some choices for the hardware implementation based on the microarchitecture proposal. We can redraw the full ETS SoC microarchitecture as shown in the following diagram:

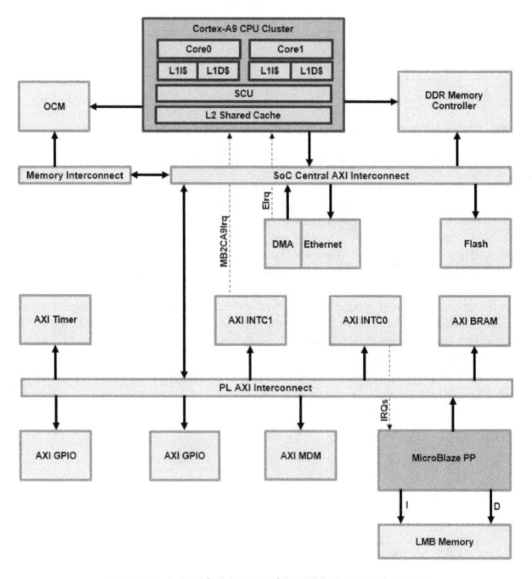

Figure 8.36 – A simplified diagram of the ETS SoC microarchitecture

The IPC interrupts from the Cortex-A9 to the MicroBlaze PP are generated using the AXI INTC0, where the doorbell registers are implemented within the AXI interrupt controller. When the Cortex-A9 needs to interrupt the MicroBlaze PP, it writes to the corresponding bit in the AXI INTC0 **Interrupt Status Register** (**ISR**), which then triggers an interrupt towards the MicroBlaze PP. Dealing with this interrupt from the MicroBlaze PP is the same as dealing with any hardware IP-generated interrupt. In the opposite direction, the process is the same – the MicroBlaze PP writes to the AXI INTC1 ISR, which then communicates through the signal output from the AXI INTC1, which is connected to the Cortex-A9 GIC input. The Cortex-A9 will deal with it as it would deal with any other hardware IP-generated interrupt.

A summary of the data exchange mechanisms for the ETS SoC Cortex-A9 and the MicroBlaze IPC

The AXI BRAM will host the circular buffer via which **Acceleration Request Entries** (**AREs**) are logged by the Cortex-A9 upon identifying an Ethernet frame for a UDP packet. The Ethernet interface uses its DMA engine to copy the received Ethernet frame from the Ethernet interface's internal buffer to the OCM memory. The Ethernet interface DMA buffer descriptors are created by the Cortex-A9 processor (at startup and before arming the Ethernet interface's DMA engine for receive operations). The DMA buffer descriptors are also created in a large circular buffer in the AXI BRAM memory – they are going to be used by the MicroBlaze PP, as it performs the filtering tasks for the Cortex-A9, so storing them in the AXI BRAM will lower the latency of their access at acceleration time. The Cortex-A9 software performs an initial frame inspection by checking the *Ethernet Type* field of the received Ethernet frame – if it finds it to be a UDP packet, it constructs an ARE data structure, which it puts in the aforementioned ARE circular buffer hosted in the AXI BRAM. When the Cortex-A9 populates the ARE circular buffer with a fresh entry, it rings the doorbell for the MicroBlaze PP by writing to the AXI INTC0 ISR, which will then trigger the corresponding interrupt toward the MicroBlaze PP. The MicroBlaze PP is the consumer of the ARE circular buffer entries, whereas the Cortex-A9 is the producer. The MicroBlaze PP maintains its read pointer (*MBARERdPtr*) of the ARE circular buffer, whereas the Cortex-A9 writes to it and maintains the write pointer (*CA9AREWrPtr*). Both pointers are visible to both processors at any time – these pointers are hosted in the AXI BRAM memory space as well. Every ARE has a recycling bit so that when the MicroBlaze PP consumes the entry and processes the request, it marks it as ready for a subsequent reuse. The following diagram from *Chapter 6, What Goes Where in a High-Speed SoC Design,* illustrates the filtering tasks offloaded to the MicroBlaze PP:

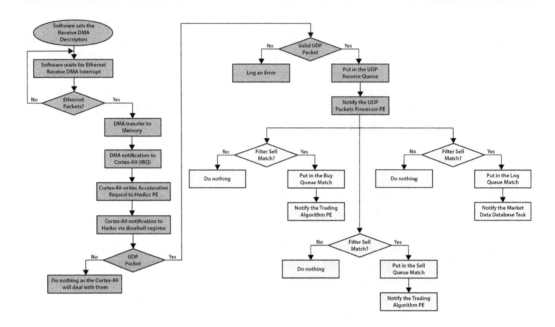

Figure 8.37 – An ETS low-latency path for a hardware-to-software interaction

When the MicroBlaze PP inspects the UDP packet associated with an ARE (there may be many UDP packets associated with a single ARE request, as we will see later), it is simply looking for a filter match. We have highlighted three filters thus far (*buy*, *sell*, and *log*). A specific UDP packet may match both the sell and log filters or both the buy and log filters. When the MicroBlaze PP finds a filter match for a specific symbol on a UDP packet, it puts its descriptor address in the corresponding response queue and rings the corresponding doorbell for the Cortex-A9. We have considered five queues in the architecture definition:

- The buy queue (*BuyQ*) that the MicroBlaze PP fills with the descriptors of the Ethernet frames carrying the UDP packet with a buy filter match on their symbol

- The sell queue (*SellQ*) that the MicroBlaze PP fills with the descriptors of the Ethernet frames carrying the UDP packet with a sell filter match on their symbol

- The market data queue (*MdataQ*) that the MicroBlaze PP fills with the descriptors of the Ethernet frames carrying the UDP packet with a market data filter match on their symbol

- The management data queue (*MgmQ*) that the MicroBlaze PP fills with the descriptors of the Ethernet frames carrying the UDP packet with a management message

- The DMA descriptor recycle queue (*DDRQ*) where the MicroBlaze PP puts the address of the descriptors of the Ethernet frames it has dealt with – this queue may seem redundant but can be used as a checking mechanism by the Cortex-A9 garbage collection tasks

Every time the MicroBlaze PP writes a descriptor in a specific queue, it rings the doorbell associated with it by sending a software-triggered interrupt to the Cortex-A9 using the AXI INTC1 mechanism.

All the queues described here are also circular buffers, for which the MicroBlaze PP is now the entry producer and the Cortex-A9 is the entry consumer. Every queue has two pointers, a write pointer owned by the MicroBlaze PP and a read pointer owned by the Cortex-A9. Both pointers are visible to both CPUs. All the filtering results queues are hosted in the AXI BRAM memory as well as the write and read pointers. The following figure provides a summary of the filtering match queues and their associated pointers:

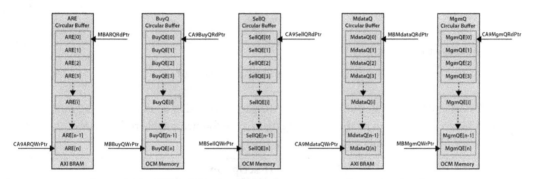

Figure 8.38 – ETS SoC filtering match data queues and associated pointers

The ETMP protocol overview

The **Electronic Trading Market Protocol** (**ETMP**) defines a single-length UDP packet payload (320 bits or 40 bytes) and has many fields, as defined in the following figure:

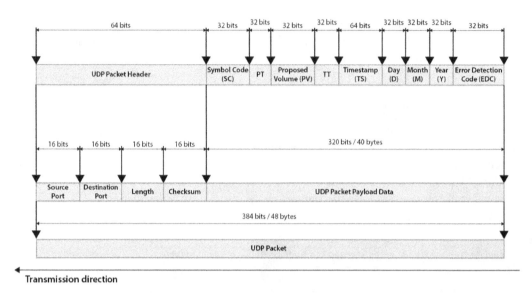

Figure 8.39 – The ETMP packet layout

The UDP header adds another 64 bits of data to the packet, resulting in an ETMP UDP frame of 384 bits or 48 bytes, as illustrated by the preceding figure. The following table reminds us of the ETMP fields that the MicroBlaze and Cortex-A9 software needs to use:

Field	Length in bits	Description
Symbol Code (SC)	32	The financial traded product symbol code. Every financial product has a unique code assigned by the ETM when the product is first introduced to the ETM.
Packet Type (PT)	32	States whether this is a Market Management packet or a Market Data packet. 0b0: Market Data packet. 0b1: Management packet.
Proposed Volume (PV)	32	The proposed maximum volume for a sell or buy action. Partial proposals of trade can be made by the ETS if interested in the symbol.
Transaction Type (TT)	32	The transaction type associated with this financial product, buying, or selling: 0b0: Buying. 0b1: Selling.

Field	Length in bits	Description
Timestamp (TS)	64	This is the timestamp logging when the UDP packet left the ETM servers.
Day (D)	32	Encodes the day when the UDP packet was sent.
Month (M)	32	Encodes the month when the UDP packet was sent.
Year (Y)	32	Encodes the year when the UDP packet was sent.
Error Detection Code (EDC)	32	CRC32 computed over all the ETMP packets excluding itself (over the 288 bits).

Table 8.1 – A description of the ETMP packet format and fields

For the EDC, there are open source implementations of the CRC32 algorithm in C that we can use for now in this design example. We will revisit this in *Part 3* of this book when we look at *profiling and hardware acceleration techniques* in detail to modify the design to include a hardware-based CRC32 implementation.

The ETS SoC system address map

The system address map allows us to locate the physical address of all the mapped devices and memories in the SoC address space, as seen from the Cortex-A9 cluster AXI interfaces and the MicroBlaze PP. This gives us an idea of how to initialize the necessary software pointers when we want to allocate their associated storage, for example, as we develop the software applications for the SoC design.

The ETS SoC MicroBlaze PP system address map

To access the MicroBlaze PP system address map, we can simply click on the Hardware mapping details in the Vitis IDE's main window. The MicroBlaze PP system address map looks as follows:

IP	Base Address	High Address	Description
LMB memory	0x0000_0000	0x0000_3FFF	SLMB
A_AXI_GPO.PS7_DDR_0	0x2000_0000	0x3FFF_FFFF	PL AXI Interconnect
AXI Timer	0x4000_0000	0x4000_FFFF	PL AXI Interconnect
MicroBlaze Debug MDM	0x4040_0000	0x4040_0FFF	PL AXI Interconnect
AXI INTC0	0x4080_0000	0x4080_FFFF	PL AXI Interconnect
AXI GPIO	0x4120_0000	0x4120_FFFF	PL AXI Interconnect
AXI BRAM memory	0x4200_0000	0x4200_3FFF	PL AXI Interconnect
AXI INTC1	0x4240_0000	0x4240_FFFF	PL AXI Interconnect

Table 8.2 – The MicroBlaze PP system address map

The ETS SoC Cortex-A9 system address map

To access the Cortex-A9 system address map, we can simply click on the hardware mapping details in the Vitis IDE's main window. The Cortex-A9 system address map looks as follows:

IP	Base Address	High Address	Description
PS7_RAM_0	0x0000_0000	0x0002_FFFF	
PS7_DDR_0	0x0010_0000	0x3FFF_FFFF	Direct port mapping
AXI GPIO	0x4120_0000	0x4120_FFFF	PL AXI Interconnect
AXI BRAM memory	0x4200_0000	0x4200_3FFF	PL AXI Interconnect
AXI INTC1	0x4240_0000	0x4240_FFFF	PL AXI Interconnect
PS7_UART_1	0xE000_1000	0xE000_1FFF	PS AXI Central Interconnect
PS7_I2C_0	0xE000_4000	0xE000_4FFF	PS AXI Central Interconnect
PS7_GPIO_0	0xE000_A000	0xE000_AFFF	PS AXI Central Interconnect
PS7_Ethernet_0	0xE000_B000	0xE000_BFFF	PS AXI Central Interconnect
PS7_QSPI_0	0xE000_D000	0xE000_DFFF	PS AXI Central Interconnect
PS7_IOP_BUS_CFG_0	0xE020_0000	0xE020_0FFF	PS AXI Central Interconnect
PS7_SLCR_0	0xF800_0000	0xF800_0FFF	Internal to the CPU Cluster
PS7_DMA_NS	0xF800_4000	0xF800_4FFF	PS AXI Central Interconnect
PS7_DMA_S	0xF800_3000	0xF800_3FFF	PS AXI Central Interconnect
PS7_DDRC_0	0xF800_6000	0xF800_6FFF	
PS7_DEV_CFG_0	0xF800_7000	0xF800_70FF	PS AXI Central Interconnect
PS7_XADC_0	0xF800_7100	0xF800_7120	PS AXI Central Interconnect
PS7_AFI_0	0xF800_8000	0xF800_8FFF	
PS7_AFI_1	0xF800_9000	0xF800_9FFF	
PS7_AFI_2	0xF800_A000	0xF800_AFFF	
PS7_AFI_3	0xF800_B000	0xF800_BFFF	
P7_OCMC_0	0xF800_C000	0xF800_CFFF	
PS7_CORESIGHT_0 (1)	0xF880_0000	0xF88F_FFFF	PS AXI Central Interconnect
PS7_PMU_0	0xF889_3000	0xF889_3FFF	
PS7_GPV_0	0xF890_7000	0xF89F_FFFF	PS AXI Central Interconnect
PS7_SCUC_0	0xF8F0_0000	0xF8F0_00FC	Internal to the CPU Cluster
PS7_SCUGICC_0	0xF8F0_0100	0xF8F0_01FF	Direct mapping
PS7_SCUTIMER_0	0xF8F0_0600	0xF8F0_061F	Internal to the CPU Cluster
PS7_GLOBALTIMER_0	0xF8F0_0200	0xF8F0_02FF	
PS7_SCUWDT_0	0xF8F0_0620	0xF8F0_06FF	Internal to the CPU Cluster

IP	Base Address	High Address	Description
PS7_INTC_DIST	0xF8F0_1000	0xF8F0_1FFF	Internal to the CPU Cluster
PS7_L2CACHEC_0	0xF8F0_2000	0xF8F0_2FFF	Internal to the CPU Cluster
PS7_QSPI_LINEAR_0	0xFC00_0000	0xFCFF_FFFF	PS AXI Central Interconnect
PS7_RAM_1	0xFFFF_0000	0xFFFF_FDFF	

Table 8.3 – The Cortex-A9 system address map

(1) CoreSight is the ARM debug infrastructure used with ARM processors.

The Ethernet MAC and its DMA engine software control mechanisms

One of the most important IPs and one of the most complex peripherals used in the ETS SoC is the Ethernet interface. It connects the ETS SoC to the ETM switch, via which the UDP packets for processing are received using its DMA engine. We need to create the DMA buffer descriptors circular buffer so the received Ethernet frames will automatically be copied to the target memory using the information provided by the DMA buffer descriptors. We have already decided that the circular buffer containing the Ethernet interface DMA buffer descriptors will be hosted in the AXI BRAM memory. This memory should be marked as non-cacheable by both processors since the SoC interconnect is non-coherent. The DMA engine may change data in the DMA buffer descriptors, whereas the processors have no way of knowing about this if they keep working on the local copy that they hold in their respective data cache. For the Ethernet frames data itself, we can target any memory within the ETS SoC as far as it is visible to both the Cortex-A9 and the MicroBlaze PP processors. From the system address maps in *Tables 8.2* and *8.3*, we can see that both the AXI BRAM and the ETS SoC DDR memory can host the Ethernet frames and we can therefore use the DDR memory for this buffering given its larger capacity. We are interfacing to the DDR memory through the General-Purpose AXI interfaces (*GP0*) since we are not expecting any challenging traffic over this path, but the optimal option would have been using the High-Performance AXI interfaces, which connect directly to the memory. We can easily change this post-deployment if we discover that there is an issue with meeting the target performance using the AXI GP interface. To use the Ethernet interface in the ETS SoC software, it needs to be initialized by the Cortex-A9 software – here are the steps required to get the Ethernet interface ready for use by the software application:

1. Unlock the *System Level Control Register* so control registers can be written by software.

2. We need to configure the clocking for the 1 Gbps operations.

3. Now, we can lock the *System Level Control Register* from the software.

4. We can now initialize the Ethernet interface using the following functions provided by the Ethernet drivers:

```
Config = XEmacPs_LookupConfig(EmacPsDeviceId);
Status = XEmacPs_CfgInitialize(EmacPsInstancePtr,
Config,Config->BaseAddress);
```

5. We can now set the *MAC Address* of the Ethernet interface using the following Ethernet driver function:

```
Status = XEmacPs_SetMacAddress(EmacPsInstancePtr,
EmacPsMAC, 1);
```

6. Now, we can set the *callback functions* to handle the send event that follows the execution of a transmission operation, a receive event that follows a receive operation, and an error event in case the Ethernet interface detects an error:

```
Status = XEmacPs_SetHandler(EmacPsInstancePtr,
                            XEMACPS_HANDLER_DMASEND,
                            (void *) XEmacPsSendHandler,
                            EmacPsInstancePtr);
Status |= XEmacPs_SetHandler(EmacPsInstancePtr,
                             XEMACPS_HANDLER_DMARECV,
                             (void *) XEmacPsRecvHandler,
                             EmacPsInstancePtr);
Status |= XEmacPs_SetHandler(EmacPsInstancePtr,
                             XEMACPS_HANDLER_ERROR,
                             (void *) XEmacPsErrorHandler,
                             EmacPsInstancePtr);
```

More steps and further details of the Ethernet interface configuration are still required for a full functional set up.

> **Information**
>
> The details of the Ethernet interface driver functions used in these code snippets are available from Xilinx at https://xilinx-wiki.atlassian.net/wiki/spaces/A/pages/18841610/AXI+Ethernet+Standalone+Driver.

Next is the Ethernet DMA operation setup for the receive side – it can also be started using the following steps:

1. Using the following BSP function, we can make the 1 MB region where the *AXI BRAM* is mapped (as seen by the Cortex-A9 processor) uncacheable:

```
// The memory is made uncacheable by writing the MMU TLB
using:
Xil_SetTlbAttributes(0x42000000, 0xc02);
```

2. We can now define the DMA buffer descriptor's circular buffer in the AXI BRAM memory using the following driver functions:

```
        XEmacPs_BdClear(&BdTemplate);
XEmacPs_BdRingCreate(&(XEmacPs_
GetRxRing(EmacPsInstancePtr)),
                     RX_BD_LIST_START_ADDRESS,
                     RX_BD_LIST_HIGH_ADDRESS, XEMACPS_BD_
ALIGNMENT,
                     RXBD_CNT);
XEmacPs_BdRingClone(&(XEmacPs_
GetRxRing(EmacPsInstancePtr)),
                         &BdTemplate, XEMACPS_RECV);
```

Once the configuration and initialization steps are performed using the Xilinx-provided Ethernet interface driver functions, the system setup can be performed. We obviously also need to set up the interrupt controller and then enable the Ethernet interface interrupts. Transmit and receive operations using the Ethernet interface can then be started by ringing the DMA doorbell.

The AXI INTC software control mechanisms

The AXI INTC is used for managing the system functional interrupts of the MicroBlaze PP, and for generating the IPC software-generated interrupts between the Cortex-A9 and the MicroBlaze processors. Xilinx Vitis generates all the necessary driver functions to configure and use the AXI INTC. The following steps in the source code list how these are used in the **Peripheral Tests** template software application:

```
// Initialize the interrupt controller driver so that it is
ready to use.
XIntc_Initialize(IntcInstancePtr, DeviceId);
// Initialize the exception table.
Xil_ExceptionInit();
```

```
// Register the interrupt controller handler with the exception
table.
Xil_ExceptionRegisterHandler(XIL_EXCEPTION_ID_INT,(Xil_
ExceptionHandler)XIntc_DeviceInterruptHandler, (void*) 0);
// Enable exceptions.
Xil_ExceptionEnable();
// Start the interrupt controller such that interrupts are
// enabled for all devices that cause interrupts.
XIntc_Start(IntcInstancePtr, XIN_REAL_MODE);
```

To trigger a software interrupt using the AXI INTC, as we have introduced in *Chapter 7, FPGA SoC Hardware Design and Verification Flow*, we simply need to write 0b1 to the corresponding bit in the ISR register. The following AXI INTC driver function can be used to achieve this:

```
XIntc_Out32(IntcBaseAddress + XIN_ISR_OFFSET, INTC_DEVICE_INT_
MASK);
```

Quantitative analysis and system performance estimation

The Ethernet frames broadcasted by the ETM are 86 bytes long. At 1 Gbps (128 MB/s), we are looking at a maximum receive rate of an Ethernet frame every 640 ns, as estimated by the following formula:

$$Rate = \frac{Data\_Size}{Throughput} = \frac{86\ B}{128\ MB}\ (s) = 640\ ns$$

The maximum rate at which the ETM can send the Ethernet frames of 86 bytes each is a frame every 640 ns. Since the PL design is running at 100 MHz, that only gives the MicroBlaze PP 64 cycles to process a UDP frame. This is impossible to meet with the current proposal. This is obviously a very high rate for the type of accelerator we have decided to use in the proposal microarchitecture. We have chosen a MicroBlaze PP as a convenient way of also learning how to use it to build an FPGA-embedded processor as a coprocessor to the Cortex-A9. To be realistic, we need the ETM Ethernet transfer rate to be much lower than sending one packet every 640 ns. Without any profiling exercise on the MicroBlaze PP software, which we haven't written yet, we can't tell for sure how many cycles the MicroBlaze PP needs to look up the fields in the ETMP packet, detect the filter matches, and then produce a result for the Cortex-A9 via the envisaged mechanisms, and send a notification via the IPC interrupts. We also have decided to use a CRC32 algorithm in software, which will only make matters worse in terms of performance, but this can easily be fixed by designing a hardware-based CRC32 calculator and adding it as a coprocessor to the MicroBlaze PP itself. When we cover profiling and hardware acceleration in *Part 3*, we will keep these considerations in mind. We estimate that performing the CRC32 computing in hardware will be at least an order of magnitude faster than using the MicroBlaze PP itself to perform it. We estimate that for a GNU-based CRC32 software calculator, we need about 16 clock cycles per byte of data – that is, for a full ETMP UDP payload of 40 bytes, it shall amount to 640 clock cycles. Using a hardware-based calculator will require about a byte per clock cycle – that is, a total of 40 clock cycles.

To estimate the lookup rate for a filter match beside the CRC32 computing, and once we propose the full software microarchitecture for the MicroBlaze PP, we should be in a better position to put some numbers to the operations to perform per received Ethernet frame, therefore allowing us to predict how many system clock cycles the MicroBlaze PP will need to perform the necessary operations.

The ETS SoC Cortex-A9 software microarchitecture

Following a power-up or a cold system boot, the Cortex-A9 will perform the following tasks in software:

- Configure all the necessary system IPs including its MMU and caches
- Create the DMA buffer descriptors and arm the Ethernet MAC DMA receive engine

We can list the tasks that need to be executed following the reception of an IPC interrupt from the MicroBlaze PP when a filter is matched for a specific UDP packet. The interrupt service routine can set a flag, which the main() function can then use as a trigger to pass execution to the corresponding function associated with it. We can obviously benefit from the services of a **Real-Time Operating System** (**RTOS**) to help with performing the scheduling and task priority management, as well as providing a TCP/IP stack. Via the stack, we can then send the TCP packets back to the ETM when a buy or sell action is the result. We can also use its filesystem and flash management services to log the data of interest. As for the management packets received from the ETM, we can pass them over the PCIe link toward the host server, which deals with the policy and adjusts the algorithms that execute the trading decisions running on the Cortex-A9 software. In this chapter, we will only focus on the acceleration path back to the Cortex-A9, whereas in *Part 3*, when we introduce the use of an RTOS with the ETS SoC, we can complete the user application using these services. The following diagram provides a software microarchitecture based on the analysis performed thus far:

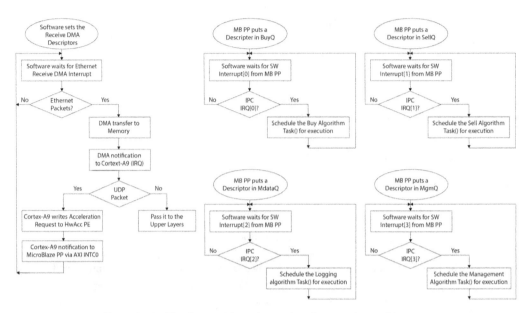

Figure 8.40 – The Cortex-A9 receive path software microarchitecture

The ETS SoC MicroBlaze PP software microarchitecture

Following a power-up or a cold system boot, the MicroBlaze PP will perform the following tasks in software:

- Configure all the necessary system IPs

- Go through the process of setting up the ISRs (ISR()) associated with the expected tasks to perform

We can list the tasks that need to be executed following the reception of an IPC interrupt from the Cortex-A9 when an ARE is received via the ARE circular buffer. The ISR function, ISR(), can set a flag, which the main() function can then use as a trigger to pass execution to the corresponding function associated with it. There will be no nested interrupt support nor filtering job preemption, so when the MicroBlaze PP detects an IPC interrupt and start executing it, it disables the interrupts and will only re-enable them upon finishing the filtering of the descriptor(s) associated with the received ARE. This task includes the generation of a response by putting the filter matching the descriptor into its destination response queue and then writing to the AXI INTC1 corresponding bit to generate the IPC interrupt toward the Cortex-A9 processor. In fact, the MicroBlaze PP, when it falls behind on the filtering and while it still has entries in the ARE circular buffer, can continue to process them until it reaches the end of the queue. When it reaches the end of the queue, it can go to sleep, pending an IPC interrupt from the Cortex-A9 for further acceleration requests. The MicroBlaze PP will need to manage the circular buffer read, *MBARERdPtr*. The following diagram provides a software microarchitecture based on the analysis performed thus far:

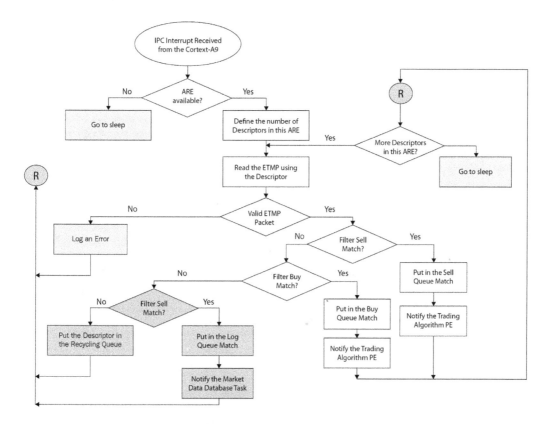

Figure 8.41 – The MicroBlaze PP software microarchitecture

Building the user software applications to initialize and test the SoC hardware

The Vitis IDE is based on Eclipse – it inherits all the source code editing features and project management Eclipse is known for. Let's explore the software project structure and how source code files can be added or removed from the project, for example:

1. In the **Vitis IDE Explorer** window, expand the src folder under one of the projects, such as ETS_SOC_CA9 – all the included sources will be listed. Double-click on the testperiph.c file to open it in the source code editor:

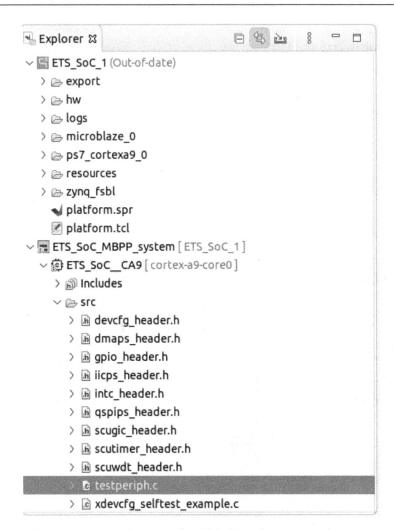

Figure 8.42 – Browsing the ETS SoC projects source code

2. The source file is now opened in the Vitis IDE as shown:

testperiph.c ⊠

```
20   * This file is a generated sample test application.
21   *
22   * This application is intended to test and/or illustrate some
23   * functionality of your system.  The contents of this file may
24   * vary depending on the IP in your system and may use existing
25   * IP driver functions.  These drivers will be generated in your
26   * SDK application project when you run the "Generate Libraries" menu item.
27   *
28   */
29
30   #include <stdio.h>
31   #include "xparameters.h"
32   #include "xil_cache.h"
33   #include "xintc.h"
34   #include "intc_header.h"
35   #include "xgpio.h"
36   #include "gpio_header.h"
37   #include "xtmrctr.h"
38   #include "tmrctr_header.h"
39   #include "tmrctr_intr_header.h"
40   int main ()
41   {
42       static XIntc intc;
43       static XTmrCtr axi_timer_0_Timer;
44       Xil_ICacheEnable();
45       Xil_DCacheEnable();
46       print("---Entering main---\n\r");
47
```

Figure 8.43 – Editing the ETS SoC project source code

Specifying the linker script for the ETS SoC projects

Once we have all the source code in place, such as for the ETS SoC design test applications of the Cortex-A9 and the MicroBlaze processors, we need to specify a linker script for each of the projects which will assign a physical location to the different sections of the executable files. The Vitis IDE has a graphical tool to edit and generate the linker script file. For both projects, follow the next steps, which will explain the linker script concept and how it can be used to assign a specific section of the executable file to a specific region of memory visible to the Cortex-A9 processor:

1. To launch the **Linker Script** editor, double-click on the **lscript.ld** entry from the **Explorer** menu as shown:

Figure 8.44 – Launching the linker script in the Vitis IDE

2.   This will open the **Linker Script** editor, which looks as follows:

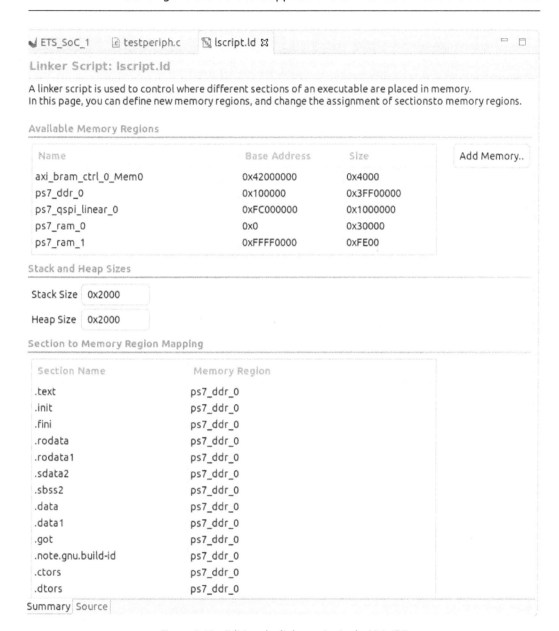

Figure 8.45 – Editing the linker script in the Vitis IDE

Setting the compilation options and building the executable file for the Cortex-A9

From the Vitis IDE, we can specify the Cortex-A9 compiler options by following these steps:

1. Right-click on the Cortex-A9-associated project in the Vitis IDE Explorer and click **Properties**:

Figure 8.46 – Accessing the build settings for the Cortex-A9 project

2. **Build Settings** will open for the project as indicated in the following figure. Under the **C/C++ Build** row, click **Settings**. Under **Arm v7 gcc assembler**, click **General**. Then, in front of **Include Paths -I**, click the + sign as shown:

Figure 8.47 – Accessing the build settings for the Cortex-A9 project

The **Add directory path** window will open as shown. Browse to where the BSP <include> directory is located on your machine. It should be under <project location path>/ ETS_SoC_1/ps_7_cortexa9_0/cortex-a9-core0/bsp/ps7_cortexa9_0/. Once located, select it and click **OK**:

Figure 8.48 – Adding the BSP <include> directory to the software project in the Vitis IDE

3. This will pass the BSP <include> directory to the **-I compiler** options. If this is not specified, the project won't build:

Figure 8.49 – The BSP <include> directory added to the -I compiler option in the Vitis IDE

4. Now, the projects can be built in the Vitis IDE. There are many ways to do this – for example, by going to the main menu and selecting **Project | Build all…**. The BSPs and all the binaries will be built, as shown by the following figure:

Figure 8.50 – Building all the ETS SoC software projects in the Vitis IDE

We have now built all the application software associated with the ETS SoC project using the Vitis IDE. This will then allow us to proceed to the next step of the design process, in which we will be looking at the hardware and software integration step, and we will cover this in the next chapter.

Summary

In this chapter, we started by exporting the ETS SoC hardware design into the Vitis IDE by generating the XSA file. We then used it in the Vitis IDE to create a custom hardware system definition for which we want to develop the application software. We have seen how a domain can be created in Vitis IDE for a given processor, and how a template application project can be generated and linked to a given domain. We then explored the BSP components and how they can be set up in the Vitis IDE for both the MicroBlaze and the Cortex-A9 processors to specify the Xilinx device drivers and the available software libraries. We then went back to the ETS SoC system architecture and we developed the software microarchitecture for both the Cortex-A9 receive path and the MicroBlaze PP acceleration software. We started doing some analytical work on the system performance and how we can compute some metrics for our ETS SoC design knowing only a few system parameters and without building the full design to measure them. We have also gained the necessary familiarity with how the software build options are performed in the Vitis IDE, including the use of the graphical interface to generate the linker script, as well as how the compiler options are specified within the Vitis IDE. We finally built the test applications linked to the ETS SoC project that we generated as templates when we first created the domains in Vitis. In this chapter, we performed all the necessary steps and gained most of the important knowledge required to be able to complete the full ETS SoC software applications building.

In the next chapter, we will complete the picture of what is specific to the FPGA SoC designs. We will be able to take the software binaries and combine them with the hardware bitstream to boot the complete SoC. We will also address all the aspects of the software and hardware integration to be able to solve any challenges that this final design phase may pose.

Questions

Answer the following questions to test your knowledge of this chapter:

1. What are the main steps that need to be performed to start building the software for the ETS SoC project?
2. What are the main options available for XSA file generation? Explain the differences between them.
3. What needs to be done to generate the boot software for the ETS SoC project when the Vitis project is first created?
4. What is a domain in the Vitis IDE, what are the steps to create one, and what is it needed for?
5. What is a BSP and how is it set up in the Vitis IDE?
6. How can we add a library to a software project in Vitis and what are the build option requirements for it to be recognized?
7. Propose a data structure format for the ARE that meets the requirements of the microarchitecture of the ETS SoC design.

8. Is the IPC interrupt from the Cortex-A9 necessary for this system architecture to work?

9. Suggested another alternative IPC mechanism that avoids the IPC interrupts from the Cortex-A9 to the MicroBlaze processor.

10. What are the pros and cons of your suggestion in comparison to the ETS SoC microarchitecture proposal of this book?

11. How can we improve the performance of the MicroBlaze PP when executing the acceleration tasks?

12. How can we set the compiler options in the Vitis IDE?

13. What are the major sections in an executable file? How do we map them to physical memory at compile time?

14. Can we use the OCM memory as a shared memory space between the Cortex-A9 and the MicroBlaze PP?

15. What is the role of the System Level Control Register?

9

SoC Design Hardware and Software Integration

In this chapter, you will complete the steps involved in an SoC-based FPGA design. You will download the FPGA binary configuration file describing the ETS SoC to the target hardware board if you have one to hand. If no demo board capable of hosting the ETS SoC hardware design is available, you will still be able to use an emulator platform based on a system model provided by Xilinx to perform some of the SoC system design integration tasks. The virtual system will allow you to boot the SoC CPU, load it with the executable file to simply run it, or proceed to debug the application software on the virtual target as if it was the real hardware hosting it. This chapter is mainly hands-on in that you will be guided through every step of the SoC hardware and software integration and testing phases using the Vitis IDE and the Xilinx emulation platform. This is the closing chapter of *Part 2*, which has covered most of the SoC system design and development topics introduced in *Part 1*, but with an emphasis on their practical aspects. We have based this chapter on a complex design involving many of the topics we have studied thus far and tried to illustrate a few of them at a time, so the principles will be mastered, and you are prepared to tackle more complex topics in both the hardware and software design techniques of modern SoCs.

We have gone from the concept of designing an Electronic Trading System (ETS) SoC to building its actual components. We have specified the design and the system requirements, built the hardware, and exported the design from the hardware design environment to the software design environment. Within the Vitis IDE, we configured the BSP and the drivers and generated the test software for both ETS SoC CPUs, and in this chapter, we will put all these parts together to run the software on the ETS SoC hardware. We will be using a real electronics board or a virtual platform emulating the SoC in its operating environment.

In this chapter, we're going to cover the following main topics:

- Connecting to an FPGA SoC board and configuring the device
- The emulation platform for running the embedded software

- Using the emulation platform to debug the SoC test software
- Embedded software profiling using the Vitis IDE

Technical requirements

The GitHub repo for this title can be found here: `https://github.com/PacktPublishing/Architecting-and-Building-High-Speed-SoCs`.

Code in Action videos for this chapter: `http://bit.ly/3fOdrEc`.

Connecting to an FPGA SoC board and configuring the FPGA

Once the application software for the ETS SoC design has been compiled using the Vitis IDE and its executable image has been generated, we can proceed to the next phase: connecting to the FPGA board from the Vitis IDE to download the FPGA SoC bitstream. This phase requires that a JTAG cable is connected from the host machine to the demo board. The JTAG cable is usually plugged into the host machine's USB port. If you are using the UbuntuVM environment, as suggested in this book, the drivers for the Xilinx JTAG cable aren't installed, and you will need to install them separately using the Vivado Lab solutions available at `https://www.xilinx.com/support/download.html`.

This will install the JTAG cable drivers and the necessary device packages that will allow this version of the Vivado tools to connect to the hardware board. Alternatively, you will need to use Vivado and Vitis IDEs on a native OS version that is officially supported by the Xilinx tools. This will then install the JTAG cable drivers as part of the Vitis installation process, and therefore you can connect directly from the Vitis IDE to the board. In the remaining parts of this chapter, we will focus on the Xilinx emulator platform to maximize the usefulness of these practical parts to everyone.

Another aspect of the hardware and software integration in the Xilinx SoC design flow is the automatic loading of the executable file that was mapped by the linker script to the FPGA BRAMs. In the ETS SoC design, all the sections of the MicroBlaze PP executable file were mapped to the LMB BRAM by the linker script. Given that the BRAMs are part of the PL side of the FPGA, they are configured from a hardware design perspective via the FPGA bitstream like any other hardware macro of the FPGA, but the Vitis tools also initialize these BRAMs' content with the executable file sections. This step is automatically run whenever a **Build All** command is executed in Vitis using a bootloop executable file to initialize the MicroBlaze LMB BRAM with. This can be changed to the test application executable we built for the MicroBlaze PP in *Chapter 7, FPGA SoC Hardware Design and Verification Flow*, as can be seen in *Step 4*.

To download the FPGA bitstream from the Vitis IDE, follow these steps:

1. From the Vitis IDE main menu, click **Xilinx | Program Device**:

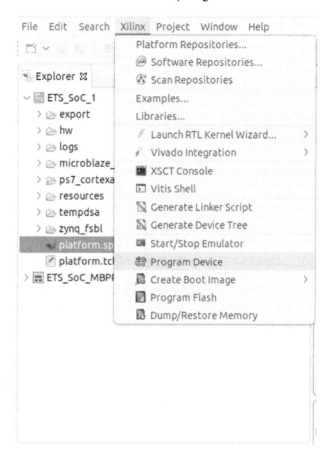

Figure 9.1 – Launching the Vitis IDE Program Device wizard

2. In the **Program Device** window shown in the following screenshot, set the bitstream file path, the **BMM/MMI File** location, and the executable file to load into MicroBlaze. Leave all the other settings as the defaults and click **Program**. This will launch the board's JTAG discovery and then program the Zynq-7000 FPGA device with the bitstream initializing the LMB BRAM with the MicroBlaze PP **bootloop** executable.

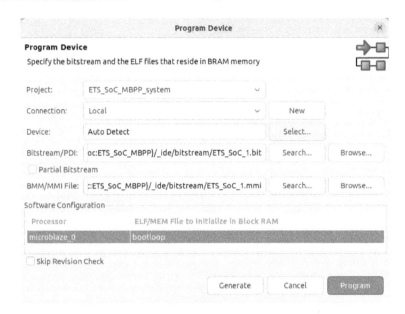

Figure 9.2 – Setting the Program Device options

3. The MicroBlaze PP executable file can be changed to the one built in *Chapter 8, FPGA SoC Software Design Flow,* for the ETS SoC project. You can modify it by browsing to its location and selecting it as the executable file for the **microblaze_0** processor instance in the **Program Device** wizard.

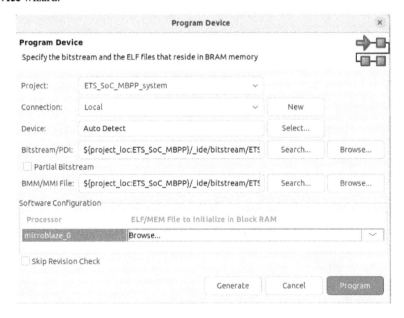

Figure 9.3 – Changing the executable file for the MicroBlaze PP

4. When you select **Browse…**, press *Enter* on your keyboard and the executable file associated with the microblaze_0 hardware instance will be automatically found, which you can then select by clicking **Open**.

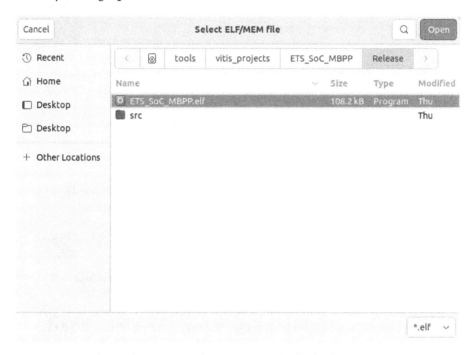

Figure 9.4 – Selecting an application executable file for the MicroBlaze PP

5. The ETS_SoC_MBPP.elf file is now mapped to the LMB BRAM in the FPGA bitstream. To update the bitstream, you can now click **Generate**, which will call the updatemem Vitis utility that updates the FPGA bitstream with the MicroBlaze PP executable **Executable and Linkable (ELF)** file.

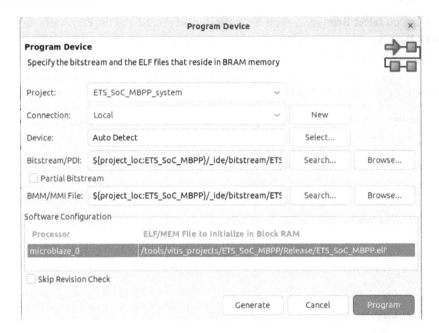

Figure 9.5 – Updating the FPGA bitstream with the MicroBlaze PP executable file

The emulation platform for running the embedded software

Xilinx provides an emulation platform for the Zynq-7000 SoC, the MicroBlaze embedded processor, and the Zynq UltraScale+ MPSoC. It is based on the **Quick Emulator** (**QEMU**), which is a Linux-based emulation platform that allows us to build a virtual system including the SoC with all its internal hardware component modules, its external interfaces, and all the surrounding electronics board integrated circuits. It is therefore a suitable environment for software development and prototyping. These ports emulate board-level systems with many peripheral models included. QEMU is already integrated within the Vitis IDE and can be directly called as if we were targeting a demo board.

> **Information**
>
> For the list of features and supported devices by the Xilinx QEMU ports, check *Chapter 2* of the *Xilinx Quick Emulator User Guide* at https://docs.xilinx.com/v/u/2020.1-English/ug1169-xilinx-qemu.

In the ETS SoC project, we will only be able to target a single CPU at a time: either the Cortex-A9 PS subsystem or the MicroBlaze PL subsystem. However, this is enough to get started and interact with the software part of the SoC design using the associated Vitis IDE tools for introducing debugging techniques. We will target the Cortex-A9 processor in this chapter. It is worth mentioning that Xilinx provides a way of performing QEMU and user logic co-simulation via remote ports, which uses a SystemC/TLM2.0 interface, which requires the PL side of the design to be built in SystemC, Verilog, or VHDL. This method could be a way of integrating custom logic built in the PL part of the Zynq-7000 with QEMU, and therefore have the MicroBlaze PP simulating in RTL and the Cortex-A9 of the PS in QEMU; however, this is outside the scope of this book.

Using QEMU in the Vitis IDE with the ETS SoC project

As already mentioned, QEMU is integrated into the Vitis IDE, which means that there is already a virtual platform that can be used as a prototyping environment for the test software we have built for the Cortex-A9 processor. This virtual environment presents itself as if it was the real hardware on which the executable file is running, and therefore can be used natively with the Cortex-A9 executable we built in *Chapter 8, FPGA SoC Software Design Flow*. To use QEMU as a demo platform in the Vitis IDE, just follow these steps:

1. To launch QEMU, go to **Xilinx** in the Vitis IDE main menu and then click **Start/Stop Emulator**:

Figure 9.6 – Launching QEMU in the Vitis IDE

2. The following wizard will open, showing the project and its associated configuration. Click **Start** to launch QEMU.

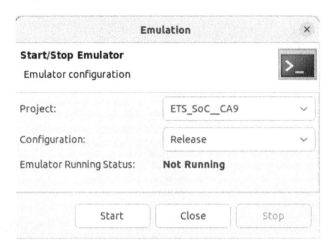

Figure 9.7 – Launch menu of QEMU in the Vitis IDE

QEMU will then start its console window in the Vitis IDE.

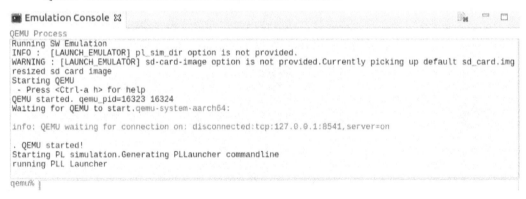

Figure 9.8 – QEMU console in the Vitis IDE

3. To load the software on the emulator and run it, click **Run** from the Vitis IDE main window.

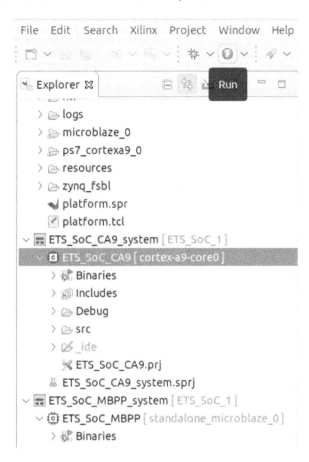

Figure 9.9 – Running the software on QEMU in the Vitis IDE

4. The following **WARNING** window will open as we have already launched a QEMU session, but without any software loaded on it. Click **OK** to relaunch QEMU, this time with the software executable of the Cortex-A9 core0 loaded on the platform and automatically started.

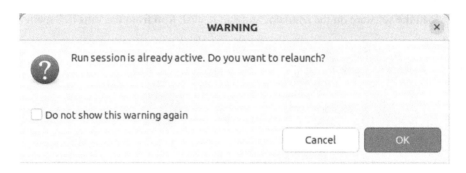

Figure 9.10 – QEMU warning in the Vitis IDE

The QEMU console will act as a terminal for the embedded software and captures its output.

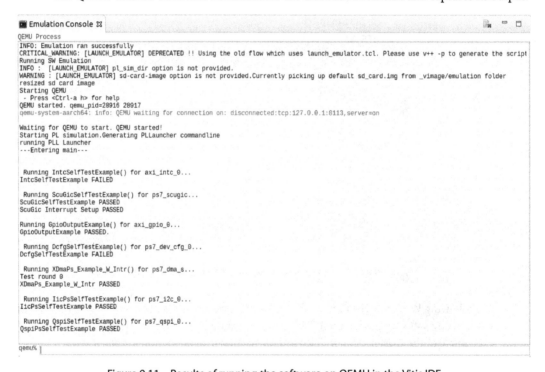

Figure 9.11 – Results of running the software on QEMU in the Vitis IDE

5.   We can observe from the results that some tests have passed and two have failed, while the last test appears to be still running. The logical action to take is to get some debugging going so we get some answers to why these didn't return PASSED. Stop the session by going to the Vitis IDE main window and clicking **Xilinx | Start/Stop Emulator**, as shown in *Figure 9.6*.

6. The following window will open. Click **Stop**.

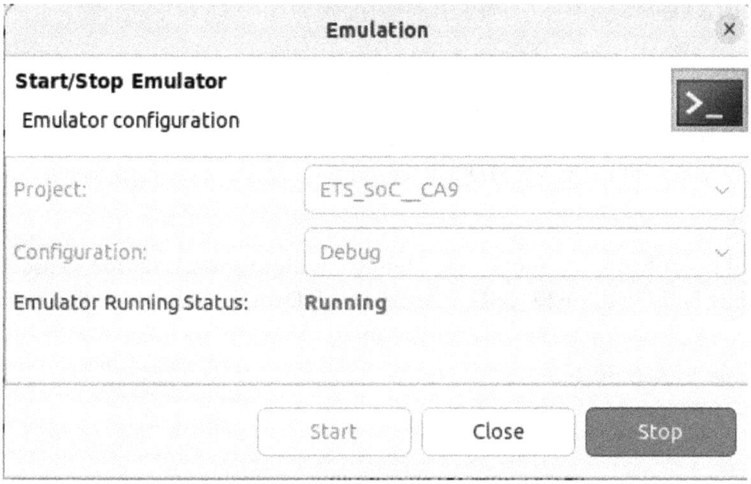

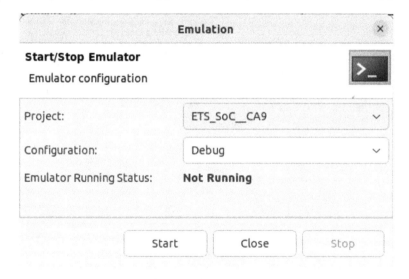

Figure 9.12 – Stopping QEMU in the Vitis IDE

The following message will then appear, confirming that QEMU has been stopped:

Figure 9.13 – QEMU session status in the Vitis IDE

Using the emulation platform for debugging the SoC test software

QEMU can be used as a virtual target to debug the software we built in *Chapter 8, FPGA SoC Software Design Flow,* for the ETS SoC project. Readers with experience of using Eclipse-based SDK debugging of embedded software running on a hardware target board will find it similar to using QEMU as a target debug environment in the Vitis IDE. To connect QEMU as a debug target, let's go through the next steps:

1.   We need to configure or examine the **Debug Configuration…** option to use. Right-click on the **Debug** build under **ETS_SoC_CA9** and click **Debug | Debug Configuration…**.

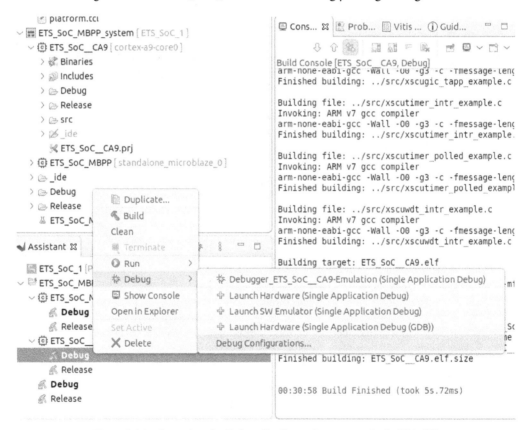

Figure 9.14 – Accessing the Debug Configuration... menu in the Vitis IDE

2. This will open the following window. Select **Debugger_ETS_SoC_CA9-Emulation**, tick the **Emulation** entry, and make sure that all the settings match the following screenshot:

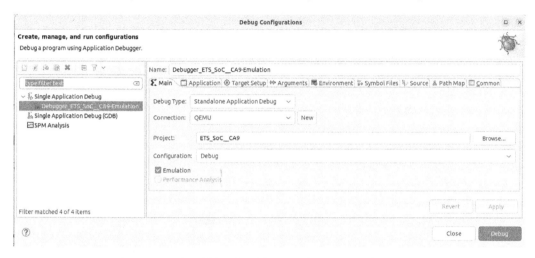

Figure 9.15 – Specifying the debug configuration in the Vitis IDE

3. Select the **Application** tab and make sure that the **Download** option specifies the executable linked to the Cortex-A9 core 0 processor (**ps7_cortexa9_0**) as shown. Make sure that **Stop at 'main'** is ticked. Click **Debug**.

Figure 9.16 – Specifying the executable to debug in the Vitis IDE

4. This will then connect the debugger to QEMU, download the executable file, and release the Cortex-A9 to run until it reaches the entry point of main(). The Vitis IDE switches its view to the Eclipse-based SDK debugger.

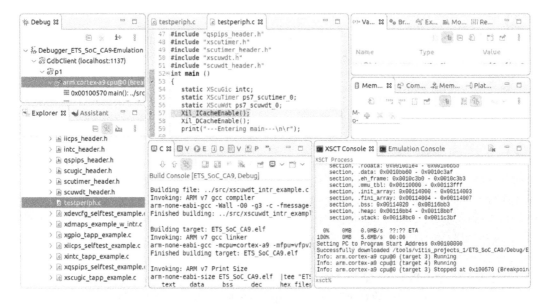

Figure 9.17 – Specifying the executable to debug in the Vitis IDE

5. Set some breakpoints and then click on **Resume**. This will run the software until it reaches the breakpoint, where it stops.

Figure 9.18 – Resuming software execution on QEMU in the Vitis IDE debugger

6.   To stop the execution of the debugger, you can click on **Suspend**.

Figure 9.19 – Suspending software execution on QEMU in the Vitis IDE debugger

7. Execution will stop and show where the software execution is at. This is in the `ps7_scuwdt_0()` test, which didn't finish executing during the free running session earlier. The Watchdog Timer test isn't hanging, it is running fine and waiting for a timeout to occur to complete the test.

Figure 9.20 – Examining the Timeout value of the Watchdog Timer model in the Vitis IDE debugger

8. To finish the debug session using QEMU, click on **Disconnect** from the Vitis IDE **Debug** view's menu.

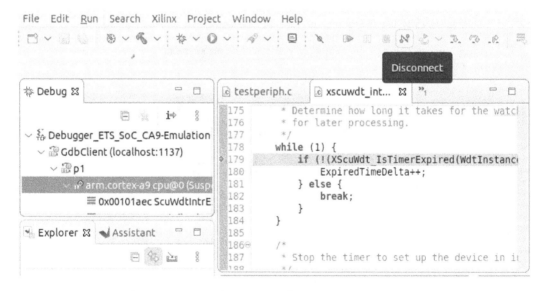

Figure 9.21 – Disconnecting the debugger from QEMU in the Vitis IDE debugger

Embedded software profiling using the Vitis IDE

To use the Vitis IDE for software profiling, we need to connect to a target hardware board. To launch the software profiling, follow the next steps:

1. Click on the arrow next to the play button in the Vitis IDE main menu. Click **Run Configurations….**

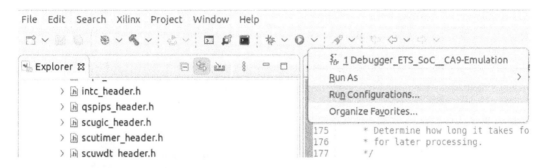

Figure 9.22 – Launching the Profiling menu in the Vitis IDE

2. The **Run Configuration** menu will open. Select the **Debugger_ETS_SoC_CA9-Performance** under **SPM Analysis** as shown. This window will allow you to configure the profiling launch options.

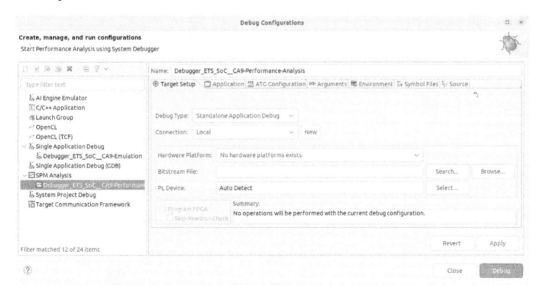

Figure 9.23 – Configuring the Profiling Launch Options in the Vitis IDE

3. Click **Debug** to connect the debugger to the target hardware board for profiling.

You can now profile the application software running on the Cortex-A9 processor on the target hardware board. You can examine the call graph and the histograms and study the system's runtime behavior to highlight areas for performance improvement.

Summary

In this chapter, we looked at the remaining final steps and utilities to complete the FPGA-based SoC system design. We have seen that using an FPGA SoC isn't so different from an SoC development targeting an ASIC technology. This is the result of the tight integration of the hardware and the software flows facilitated by the Vitis IDE, which takes over from the Vivado IDE using the XSA hardware information. The Vitis tools then inherit all the information related to the hardware platform, such as the memory map, the list of IPs, and the FPGA configuration file to boot the FPGA with, from the Vitis IDE. We have seen how easy it is to proceed to download the FPGA binary configuration file describing the ETS SoC design to the target hardware board from the Vitis IDE. Software development can start using QEMU as a virtual development platform. This emulator is easily accessible from within the Vitis IDE. The virtual system allowed us to boot the SoC CPU. We could also load it with the executable file and fully run it on the emulation target. We have also been able to perform software debugging on QEMU using an Eclipse-based SDK-like IDE debugger within the Vitis IDE. In this chapter, we have completed the last phase of the system co-design by performing the hardware and software integration. We have also executed some system-level testing using the template software test examples.

This chapter closes *Part 2* of this book, which has covered most of the SoC system design and development topics in practical, easy-to-follow steps. We have gone from the concept of designing an ETS SoC to building its actual components. We specified the design and the system requirements, built the hardware, exported the design from the hardware design environment to the software design environment, configured the BSP and the drivers, and generated the test software for both ETS SoC CPUs. We have also put all these parts together to run the software on the ETS SoC hardware using a real electronics board or a virtual platform emulating the SoC. We are now ready to tackle more complex embedded system design topics, such as platforms design for an embedded Linux or a **Real-Time Operating System** (**RTOS**) including the system-level security aspects and data coherency between the PS and the PL accelerators over the Cortex-A9 Acceleration Coherency Port. These and other FPGA SoC-based applications in different industry verticals will be discussed in *Part 3* of this book.

In the next chapter, we will explore some advanced and challenging SoC design topics, which will allow us to efficiently connect hardware accelerators to the Cortex-A9 processor using coherent data-sharing mechanisms. We will also learn how to analyze the performance of these IPC mechanisms.

Questions

Answer the following questions to test your knowledge of this chapter:

1. Is it possible to program the FPGA using the Vitis IDE? How would you do that?

2. If the MicroBlaze executable is hosted in the LMB BRAM, which utility is used to integrate it into the FPGA bitstream file?

3. Why do we need to initialize the LMB BRAM with the MicroBlaze executable prior to configuring the FPGA SoC?

4. Describe the steps used in the Vitis IDE to update the FPGA SoC bitstream using a new MicroBlaze executable ELF file.

5. What is QEMU?

6. What are the characteristics of the Xilinx QEMU? Which processors from Xilinx SoCs does it support?

7. Describe the steps to launch QEMU from the Vitis IDE.

8. How different is debugging the Cortex-A9 software using the Xilinx port of QEMU in comparison to using a real hardware board?

9. Describe the steps needed to debug the Cortex-A9 software on the QEMU emulator.

10. How can we profile the Cortex-A9 software?

Part 3: Implementation and Integration of Advanced High-Speed FPGA SoCs

This part introduces more advanced and challenging topics of the SoC and FPGA design integration, the available tools, and their criteria of selection, as well as implementing advanced systems.

This part comprises the following chapters:

- *Chapter 10, Building a Complex SoC Hardware Targeting an FPGA*
- *Chapter 11, Addressing the Security Aspects of an FPGA-Based SoC*
- *Chapter 12, Building a Complex Software with an Embedded Operating System Flow*
- *Chapter 13, Video, Image, and DSP Processing Principles in FPGAs and SoCs*
- *Chapter 14, Communication and Control Systems Implementation in FPGAs and SoCs*

Building a Complex SoC Hardware Targeting an FPGA

In this opening chapter of *Part 3*, you will be introduced to some of the advanced topics of SoC design that present many challenges to design engineers given their multidimensional nature. It will continue with the same practical approach as *Part 2* of this book by adding more complex elements to the hardware design. The hardware will also be built to be able to host an embedded **Operating System (OS)**. You will learn how to use advanced hardware acceleration techniques to help augment the system performance, and you will be equipped with the required fundamental knowledge that makes this design step less challenging. You will examine the different ways these techniques can be applied at the system level, and what aspects need considering at the architectural step in the shared data paradigm.

In this chapter, we're going to cover the following main topics:

- Building a complex SoC subsystem using the Vivado IDE
- System performance analysis and the system quantitative studies
- Addressing the system coherency and using the Cortex-A9 ACP

Technical requirements

The GitHub repo for this title can be found here: `https://github.com/PacktPublishing/Architecting-and-Building-High-Speed-SoCs`.

Code in Action videos for this chapter: `http://bit.ly/3TlFU1I`.

Building a complex SoC subsystem using Vivado IDE

We will start from the **Electronic Trading System (ETS)** SoC hardware design and add more features to it, such as connecting a master interface from within the **Programmable Logic (PL)** side of the FPGA SoC to the **Accelerator Coherency Port (ACP)** of the **Processing Subsystem (PS)** side. We will also make sure that the PS design includes all the hardware features necessary to run an embedded OS such as the timers, the storage devices, the **Input/Output (IO)** peripherals, and the communication interfaces. Let's get started:

1. Launch the Vivado IDE and open the ETS SoC hardware design we built in *Chapter 7, FPGA SoC Hardware Design and Verification Flow*.

2. Go to IP Integrator and click **Open Block Design**, causing the block design to open in the Vivado IP Integrator window.

3. Double-click on **ZYNQ7 Processing System** to open it for customization. The following window shall open:

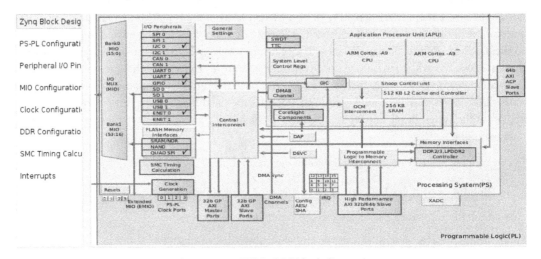

Figure 10.1 – ETS SoC PS block diagram

4. Double-click on **64b AXI ACP Slave Ports** in the ETS SoC PS block to access the ACP port configuration interface. This will open the **PS-PL Configuration** window shown in the following screenshot. In the **ACP Slave AXI Interface** row, check both rows associated with the ACP port configuration. We will cover the ACP details later in this chapter.

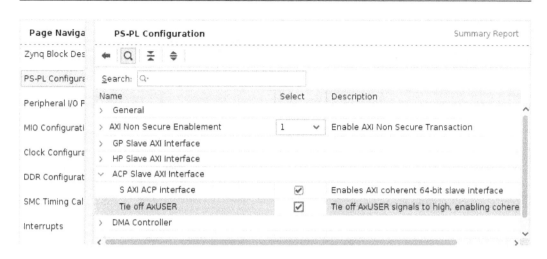

Figure 10.2 – Configuring the ACP slave AXI interface in the Vivado IDE

5. Expand the **General** row and check the **Include ACP transaction checker** option, as shown in the following figure:

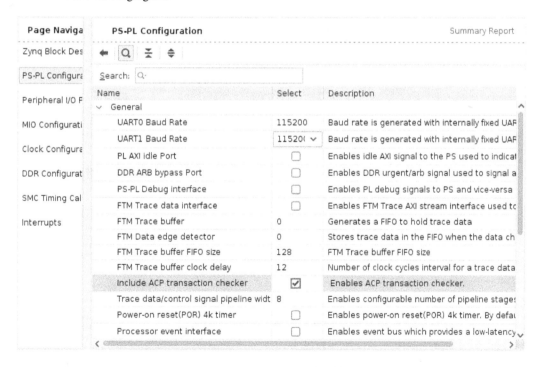

Figure 10.3 – Enabling the ACP transaction checker for the ACP port

6. We need to connect the ACP slave AXI port to the MicroBlaze subsystem interconnect. This step is performed semi-automatically by clicking on the IP Integrator **Run Connection Automation** hyperlink:

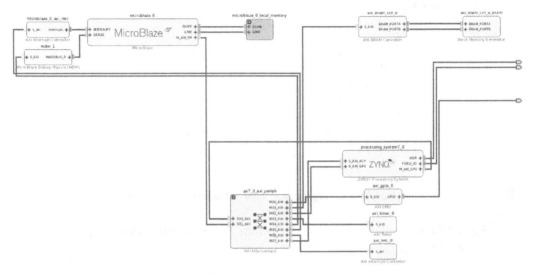

Figure 10.4 – Using Run Connection Automation to connect the ACP port

7. This will open the following window. Click **OK**:

Figure 10.5 – ACP slave port connection wizard

8. We need to add the SD IO peripheral to the PS subsystem. Double-click on **Zynq Block Design** in the **IP Navigator** pane to open the **ZYNQ7 Processing System** diagram, as shown in *Figure 10.6*.

9. Double-click in the **SD 0** row in the **IO Peripherals** column to add the SD card to the PS system. This may be used to host the filesystem of the OS if needed later by the advanced software development flow.

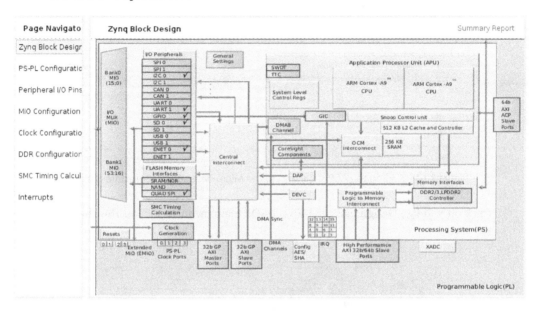

Figure 10.6 – Adding the SD IO to the PS subsystem in the Vivado IDE

10. We have included everything we need in the ETS SoC hardware to study some advanced features of the SoC hardware design flow, as well as preparing the ETS SoC hardware to host an embedded OS. We need to make sure that the ACP port is mapped and visible within the MicroBlaze address map so the MicroBlaze CPU can initiate transactions through the PS ACP port.

11. Open the address map by clicking the **Address Map** tab in the **IP Integrator** window. The system address map should open. Expand all the entries in it and make sure that you include and assign the ACP region, as shown:

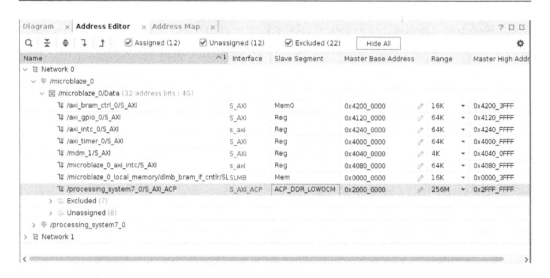

Figure 10.7 – Adding the ACP slave port to the MicroBlaze subsystem address map

12. Click on **Validate Design** using the Vivado IDE main menu, as shown. This should not raise any errors:

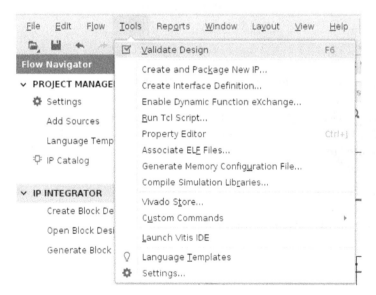

Figure 10.8 – Validating the ETS SoC design in Vivado IDE

System performance analysis and the system quantitative studies

To perform the SoC system performance analysis and quantitively study it, we need to refer to *Chapter 8, FPGA SoC Software Design Flow*, specifically, all the details from the section titled *Defining the distributed software microarchitecture for the ETS SoC processors*. We have provided a full ETS SoC microarchitecture, as shown in the following diagram:

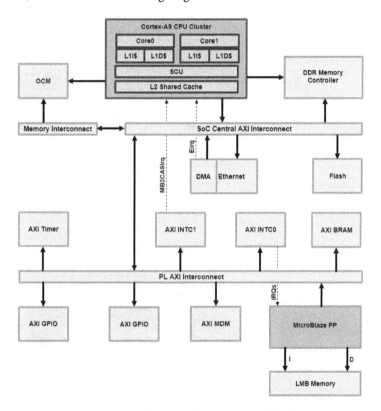

Figure 10.9 – ETS SoC microarchitecture simplified diagram

In this analysis, we aim to understand whether the proposed IPC between the Cortex-A9 CPU and the MicroBlaze PP processor mechanism is optimal and using all the possible hardware capabilities of the SoC FPGAs. We would like to figure out whether using the ACP would be a better alternative to using the microarchitecture proposal implementation via the PS AXI GP interface. The current IPC mechanism from the Cortex-A9 toward the MicroBlaze PP uses the circular buffer queue hosted in the AXI BRAM, where the **Acceleration Requests (AREs)** are built by the Cortex-A9 first and then written as entries into the ARE circular buffer. This is followed by a notification from the Cortex-A9 to the MicroBlaze PP as an interrupt by writing to the AXI INTC0. Once the MicroBlaze PP reaches the ARE entry, it uses the pointers in it to retrieve the Ethernet frame to filter from where the Ethernet

DMA engine has written it, and these are in the OCM memory. The following figure summarizes this interaction with the sequencing of events details:

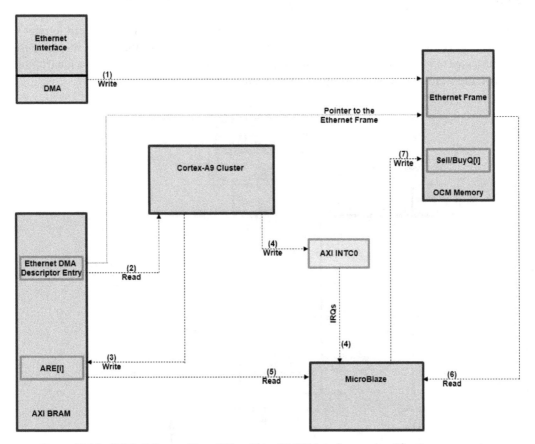

Figure 10.10 – ETS SoC Cortex-A9 and MicroBlaze PP IPC data flow and notification sequencing

The ETS SoC Cortex-A9 and MicroBlaze PP IPC data flow and the notification sequencing include the following steps:

1. When the Ethernet DMA engine receives an Ethernet frame, it uses the details provided by the Cortex-A9 CPU to copy the payload data of the Ethernet frame from the Ethernet controller buffer into the OCM memory. This operation to read the descriptor entry from the AXI BRAM by the DMA engine takes 6 **Clock Cycles (CCs)** of the PS interconnect clock (ps_clk), which is [6 x CC (ps_clk)].

2. The Cortex-A9 receives a notification from the Ethernet controller DMA engine for the received Ethernet frame. It reads the Ethernet DMA descriptor entry and uses the OCM memory pointer of the Ethernet frame to inspect its type. The preceding flow ignores the Cortex-A9 partial access to the Ethernet frame in the OCM memory as it is considered negligible and not affecting this specific analysis' outcome. This may take 8 CCs of the MicroBlaze PP system clock (pl_clk), which is [8 x CC (pl_clk)].

3. When the Cortex-A9 finds that the Ethernet frame is for a UDP packet, it constructs an ARE[i] for it in the AXI BRAM. This processing is internal within the Cortex-A9 core and may take n CCs of the Cortex-A9 clock, which is [n x CC (ca9_clk)].

4. The Cortex-A9 notifies the MicroBlaze by writing to the AXI INTC0 (AXI INTC is the Xilinx AXI interrupt controller, which combines several interrupts into a single interrupt to notify the MicroBlaze processor), which generates an interrupt for the MicroBlaze PP. This is an access to a register within the AXI INTC0 running at the MicroBlaze system clock and may take in the region of 6 CCs MicroBlaze system clock, which is [6 x CC (pl_clk)].

5. Once the MicroBlaze PP reaches the ARE[i], it reads from the AXI BRAM to retrieve the pointer to use to read the important fields of the Ethernet frame from the OCM memory. It may take in the region of 4 CCs MicroBlaze system clock, which is [4 x CC (pl_clk)].

6. The MicroBlaze PP now reads the fields of interest using the information in the ARE[i]. This will be performed a field at a time followed by the necessary computation to perform the requested filtering. If we only consider the access time of the data, the cumulative time to access it would be approximately 6 CCs per word of 32 bits each; in total, we have 12 words, which results in [72 x CC (pl_clk)].

7. Once the results are ready, the MicroBlaze PP will create an entry in the appropriate queue in the OCM and/or the AXI BRAM, and this will take approximately 6 CCs per word of 32 bits each, which is [6 x CC (pl_clk)].

In the preceding approximation, we observe that the time it takes for the IPC communication associated with a single Ethernet frame from its reception by the Ethernet DMA to providing the filtering results is the sum of all the estimated segment's times:

$$IPC_{Time} = [n \times CC\ (ca9_{clk})] + [6 \times CC\ (ps_{clk})] + [96 \times CC\ (pl_{clk})]$$

Obviously, this only includes the IPC times, which is interleaved by the packet inspection time by the MicroBlaze; however, the filtering time is another performance metric we can estimate, but it won't affect the decision to move into using the ACP in our case.

In the ETS SoC design, we have the following:

- pl_clk = 100 MHz

- ps_clk = 222 MHz

- ca9_clk = 666 MHz

The estimated IPC required time is therefore as follows:

$$IPC_{Time} = [n \times CC\ (ca9_{clk})] + [6 \times 4.5\ ns] + [96 \times 10\ ns]$$

$$IPC_{Time} = [987\ ns] + [n \times CC\ (ca9\_clk)]$$

Since *[n x CC (ca9_clk)]* is roughly the same even if we modify the IPC mechanism used, we can then use the preceding result as a base figure to compare against.

Another point worth highlighting here is that the IPC mechanism dictates the use of non-cacheable memory since the system interconnect isn't coherent within the PS nor between the PS and the PL. If cacheable memory regions are used for any of the previously involved data in the IPC mechanism, cache management operations should be carefully used to flash the Cortex-A9 data cache whenever a descriptor-related field is updated or an ARE is constructed. The use of cache management is fine but will have a hit on the Cortex-A9 CPU performance. We can study this case as an exercise, but we will need to look at the resulting assembly language instructions to be able to estimate the amount of time needed by using the cache management instructions. This is easily done using the emulation platform and running code on it that allows us to view its disassembly associated instructions.

Addressing the system coherency and using the Cortex-A9 ACP port

In this chapter, the focus for the hardware acceleration is to find ways to closely integrate the PL logic hardware accelerator with the Cortex-A9 cluster and build a more direct path between software and the acceleration hardware. This direct path should be without paying the penalty of using non-cacheable memory for data shared between the Cortex-A9 and its hardware accelerators. Using cacheable memory without any cache coherency support from the hardware imposes some performance penalties. Such sharing requires some form of synchronization between the Cortex-A9 and the PL Accelerators; for example, by using cache maintenance operation in the Cortex-A9 software. This is required following every update to the common data variables between the Cortex-A9 software and the PL hardware accelerator. The way this close integration can be achieved in the Zynq-7000 SoC is through the ACP, which provides a direct coherent path from the PL logic accelerator to the Cortex-A9 caches and doesn't require the Cortex-A9 to use any cache maintenance operations following access to a shared data variable with the PL hardware accelerator. The following diagram provides an overview of the envisaged topology to connect the MicroBlaze PP-based PL hardware accelerator to the PS:

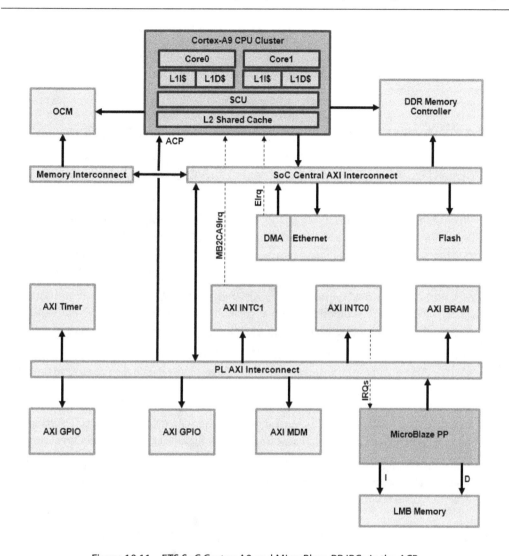

Figure 10.11 – ETS SoC Cortex-A9 and MicroBlaze PP IPC via the ACP

Overview of the Cortex-A9 CPU ACP in the Zynq-7000 SoC FPGA

The ACP is an AXI 64-bit slave port on the PS that allows the PL implemented and mapped masters to access the Cortex-A9 CPU cluster L2 and the OCM memory via transactions that are coherent with the L1 data caches of the Cortex-A9 cores and the L2 common cache. This is possible as the ACP is connected to the Cortex-A9 CPU cluster **Snoop Control Unit (SCU)**. Within the SCU, address filtering is implemented by default that will route transactions targeting the upper 1 MB or the lower 1 MB of the 4 GB system address space to the OCM memory, whereas the remaining addresses are routed to the L2 cache controller.

> **Information**
>
> For more information on the SCU address filtering, please consult *Chapter 29* of the *Zynq-7000 SoC Technical Reference Manual*: https://docs.xilinx.com/v/u/en-US/ug585-Zynq-7000-TRM.

The ACP write **Issuing Capability (IC)** is three transactions, and its **read IC** is seven transactions.

The ACP read or write requests can be coherent or non-coherent depending on the setting of the AXI AxUSER[0] and AxCACHE[1] signals, where *x* can be either **R** for **read** or **W** for **write** transactions. We distinguish the following request types:

- **ACP coherent read requests**: When ARUSER[0]=1 and ARCACHE[1]=1, and ARVALID is HIGH. This qualifies the read operation as coherent, and the data is provided directly from the processor cache holding it and returned to the ACP port. When there is no data in any of the cluster caches, the read request is forwarded further to one of the SCU AXI master ports.

- **ACP non-coherent read requests**: When ARUSER[0]=0 or ARCACHE[1]=0, and ARVALID is HIGH. This doesn't qualify the request as coherent, and the read request is forwarded to one of the available SCU AXI master ports without looking up the CPU cluster caches.

- **ACP coherent write requests**: When AWUSER[0]=1 and AWCACHE[1]=1, and AWVALID is HIGH. This qualifies the write operation as coherent. The data, when present in one of the Cortex-A9 caches, is first cleaned and invalidated from the caches. However, if the data is not cached in any of the Cortex-A9 processors, or the cache line holding it has been cleaned and invalidated, the write request is forwarded to one of the SCU AXI master ports. The transaction can be a "write allocate" into the L2 cache if specified by the qualifying signals on the AXI interface.

- **ACP non-coherent write requests**: When AWUSER[0]=0 or AWCACHE[1]=0, and AWVALID is HIGH. The transaction is not coherent, and the write request is forwarded directly to the SCU AXI master ports.

Implications of using the ACP interface in the ETS SoC design

As already introduced, the ACP interface will allow the use of a cacheable memory to store the data variables shared between the Cortex-A9 and the MicroBlaze PP to implement the IPC data flow and its associated notifications. It is worth noting that the coherency is only in one direction; that is, if the PL master were caching any data, the Cortex-A9 would have no way of knowing whether it has changed without the PL master explicitly notifying it. This is because the AXI bus is not a coherent interconnect. It is only because we are routing transactions first through the Cortex-A9 SCU that we have a way of fulfilling coherent transactions by the Cortex-A9 caches. We can now move all the content of the AXI BRAM to the OCM or DRAM memory and access them coherently through the ACP port of the Cortex-A9. The sequencing diagram showing the different data mappings becomes like the following:

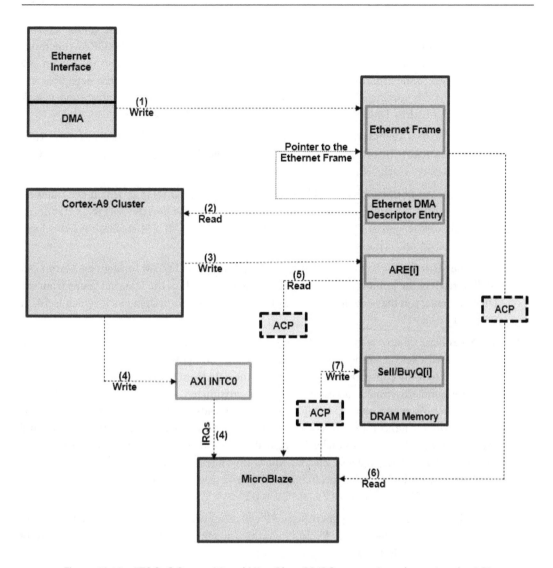

Figure 10.12 – ETS SoC Cortex-A9 and MicroBlaze PP IPC sequencing when using the ACP

Let's now quantify the ETS SoC Cortex-A9 and MicroBlaze PP IPC data flow and the notification sequencing when using the ACP. We will follow the same method we already established earlier. The sequence includes the following steps:

1. When the Ethernet DMA engine receives an Ethernet frame, it uses the details provided by the Cortex-A9 CPU to copy the payload data of the Ethernet frame from the Ethernet controller buffer into the DRAM memory. This operation to read the descriptor entry from the DRAM by the DMA engine takes 4 CCs of the PS interconnect clock (ps_clk), which is [4 x CC (ps_clk)].

2. The Cortex-A9 receives a notification from the Ethernet controller DMA engine for the received Ethernet frame. It reads the Ethernet DMA descriptor entry and uses the DRAM memory pointer of the Ethernet frame to inspect its type. The preceding flow ignores the Cortex-A9 partial access to the Ethernet frame in the DRAM memory as it is considered negligible and not affecting this specific analysis outcome. This may take 4 CCs of the PS interconnect clock (ps_clk), which is [4 x CC (ps_clk)].

3. When the Cortex-A9 finds that the Ethernet frame is for a UDP packet, it constructs an ARE[i] for it in the DRAM. This processing is internal within the Cortex-A9 core and may take n CCs of the Cortex-A9 clock, which is [n x CC (ca9_clk)].

4. The Cortex-A9 notifies the MicroBlaze by writing to the AXI INTC0, which generates an interrupt for the MicroBlaze PP. This is an access to a register within the AXI INTC0 running at the MicroBlaze system clock. It may take in the region of 6 CCs MicroBlaze system clock, which is [6 x CC (pl_clk)].

5. Once the MicroBlaze PP reaches the ARE[i], it reads from the DRAM through the Cortex-A9 ACP to retrieve the pointer to use to read the important fields of the Ethernet frame from the DRAM memory. It may take in the region of 4 CCs MicroBlaze system clock, which is [4 x CC (pl_clk)].

6. The MicroBlaze PP now reads the fields of interest using the information in the ARE[i]. This will be performed a field at a time followed by the necessary computation to perform the requested filtering. If we only consider the access time of the data from the DRAM via the Cortex-A9 ACP, the cumulative time to access it would be approximately 4 CCs per word of 32 bits each, but the MicroBlaze can issue up to seven consecutive reads before the first read completion is received. However, this is only possible if we are using a data cache on the MicroBlaze PP and therefore, we can issue up to a cache line read at a time (32 bytes). Since we haven't configured the MicroBlaze PP with a data cache, we will assume a read capability of two words at a time, and therefore, the latency is halved. In total, we have 12 words, which results in [(12 x 4/2) x CC (pl_clk)], which is [24 x CC (pl_clk)].

7. Once the results are ready, the MicroBlaze PP will create an entry in the appropriate queue in the DRAM through the Cortex-A9 ACP, and this will take approximately 4 CCs per word of 32 bits each, which is [4 x CC (pl_clk)].

In the preceding approximation, when using the ACP, we observe that the time it takes for the IPC communication associated with a single Ethernet frame from its reception by the Ethernet DMA to providing the filtering results is the sum of all the estimated segment's times:

$$IPC_{Time} = [n \; x \; CC \; (ca9_{clk})] + [8 \; x \; CC \; (ps_{clk})] + [38 \; x \; CC \; (pl_{clk})]$$

Obviously, this only includes the IPC times, which are interleaved by the packet inspection time by the MicroBlaze.

In the ETS SoC design, we have the following:

- pl_clk = 100 MHz

- ps_clk = 222 MHz

- ca9_clk = 666 MHz

The estimated IPC required time is therefore as follows:

$$IPC_{Time} = [n \; x \; CC \; (ca9_{clk})] + [8 \; x \; 4.5 \; ns] + [38 \; x \; 10 \; ns]$$

$$IPC_{Time} = [416 \; ns] + [n \; x \; CC \; (ca9\_clk)]$$

Since *[n x CC (ca9_clk)]* is roughly the same even if we have modified the IPC mechanism used, we can then use the preceding result as a base figure to compare against the previously computed one.

We observe that we have improved the IPC time by 571 nanoseconds. We are also able to use shared SoC memories cacheable by the Cortex-A9 and without the need for cache management operations to maintain the data coherency with the MicroBlaze PP.

In general, using the ACP is beneficial in system designs such as for packet processing, where the IPC latency and caching improve the application performance greatly.

Summary

In this chapter, we added a few hardware elements to the ETS SoC design to prepare it for hosting an embedded OS and improved the IPC communication between the Cortex-A9 CPU and the MicroBlaze PP. We also delved into the system performance analysis by first providing a detailed sequencing diagram of the IPC mechanism and then using it as a base to perform a quantitative study. We have used time estimates to measure how long the IPC communication associated with a received Ethernet frame to filter by the PL logic would cost. We found that a significant amount of time is needed to provide the information for moving the data and its associated descriptors from the PS domain to the PL domain. We studied the case of the IPC mechanism when using the PS AXI GP port and then studied the alternative solution of using the ACP port of the Cortex-A9. We have also exposed the issues of using cacheable memory in these scenarios and how this will require cache management operations when not using the ACP to keep the PS and PL looking and using the same copy of the shared data all the time. We computed that the use of the ACP with its coherency capability greatly reduces the time consumed by the IPC mechanism. We have also looked at most of the technical details of ACP to provide a good overview of its features and supported transactions.

In the next chapter, we will explore more advanced topics of the FPGA-based SoCs, with a focus on security.

Questions

Answer the following questions to test your knowledge of this chapter:

1. What are the main features we have added to the ETS SoC design in this chapter and why did we add them?

2. Describe the main steps needed to connect the MicroBlaze subsystem to the PS block in Vivado.

3. What modifications are needed to the address map and why?

4. Which type of transactions are supported by the ACP port?

5. List the different steps involved in the Cortex-A9 to the MicroBlaze PP IPC when using the PS AXI GP.

6. List the different steps involved in the Cortex-A9 to the MicroBlaze PP IPC when using the PS ACP.

7. How does the ACP improve the Cortex-A9 to the MicroBlaze PP IPC performance?

8. Describe a scenario (not using the ACP) when cache management operations are needed to keep the data shared between the Cortex-A9 and the MicroBlaze PP coherent.

9. List some of the disadvantages of using the ACP in general as a gateway between the PL accelerators and the Cortex-A9 memory space.

10. Why is using the ACP the best method for sharing data between the Cortex-A9 and the PL accelerators in packet processing applications such as the ETS SoC?

11

Addressing the Security Aspects of an FPGA-Based SoC

In this chapter, you will be introduced to the SoC security aspects and how they are addressed by the FPGA SoC hardware. Here, you will learn about the security paradigms available in the ARM-based processors within the SoC. You will also be introduced to the security aspects related to the software and how they make use of the hardware security features to build a secure SoC in an FPGA. This chapter will also introduce all the security features that have been specifically built by Xilinx for their FPGAs since they require a bitstream file to be stored for configuration at boot time; this is how the process is protected.

In this chapter, we're going to cover the following topics:

- FPGA SoC hardware security features
- ARM CPUs and their hardware security paradigm
- Software security aspects and how they integrate the hardware's available features
- Building a secure FPGA-based SoC

FPGA SoC hardware security features

As for any modern embedded system, connected to the internet or not, security is becoming a major design challenge, specifically in today's emerging **Internet of Things (IoT)** devices and modules. FPGA-based SoCs face the same security challenges and will need to be designed to counter illegal access and tampering. Xilinx FPGA SoCs adopt the ARM TrustZone security architecture for both the **Processing Subsystem (PS)** and **Programmable Logic (PL)** parts. The ARM TrustZone architecture is a combination of hardware and software frameworks that work in tandem to make the SoC implementing them as secure as possible. In addition to the ARM TrustZone support, Xilinx SoCs add a third dimension to the security paradigm, specifically for the PL, which requires an externally hosted configuration bitstream file. The configuration bitstream can be encrypted by the Xilinx hardware design tools; the FPGA device provides the mechanism for its decryption by the hardware before the logic is programmed with it.

The Zynq-7000 SoC FPGAs are capable of booting in a secure mode by authenticating the encrypted PS firmware images and the encrypted PL configuration bitstream. The following diagram provides an overview of the secure booting infrastructure, which shows both the PS firmware and the PL bitstream loading paths:

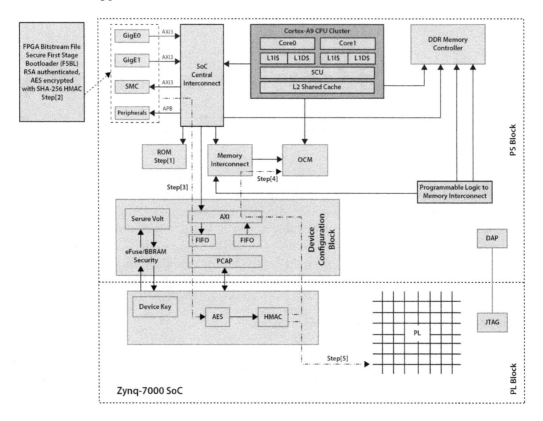

Figure 11.1 – Zynq-7000 SoC dual secure booting

The preceding diagram illustrates the secure booting steps that are executed at powerup by the Zynq-7000 SoC FPGA:

1.   First, the Zynq-7000 SoC FPGA is powered up, in which the Cortex-A9 starts executing the BootROM code.

2.   **Rivest Shamir Adleman (RSA)** authentication is performed by the Cortex-A9 BootROM code on the encrypted **First Stage Boot Loader (FSBL)**. This step is optional.

3.   FSBL **Advanced Encryption Standards (AES)** decryption and **Hash Message Authentication Code (HMAC)** authentication are performed using the SoC device hardware accelerator's AES and HMAC blocks, as shown in *Figure 11.1*.

4. The decrypted and authenticated FSBL is stored in the OCM memory within the PS block.

5. The PL (that is, the FPGA) is configured. This step is optional as it may be performed using the JTAG interface.

During Zynq-7000 SoC secure booting, many parts of the SoC device are involved. This process is always started from the SoC BootROM code. When the optional RSA authentication is enabled, the public key is used by BootROM to authenticate the FSBL before it's decrypted or executed by the Cortex-A9 processor. A chip secure boot is usually indicated by the boot image header. When this is the case, the BootROM code enables the AES and HMAC hard macro security engines, which are in the PL part of the SoC. The encrypted FSBL is copied to the AES and HMAC through the **Processor Configuration Access Port** (**PCAP**). The PCAP allows the Cortex-A9 to configure the FPGA logic when it is part of the booting process. When the FSBL image is decrypted, it is copied back through the PCAP to the OCM. Consequently, the Cortex-A9 is capable of securely configuring the FPGA logic by first decrypting the encrypted bitstream using the AES and HMAC hard macro engines via the PCAP, and then loading the FPGA configuration memory with the authenticated and decrypted configuration bitstream binary. The following flowchart illustrates the Zynq-7000 SoC boot options:

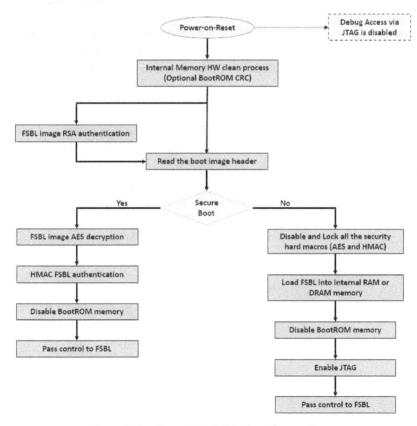

Figure 11.2 – Zynq-7000 SoC PS boot flow options

At **Power-on-Reset (PoR)**, the BootROM code is executed by the Cortex-A9. The boot options can be specified using the Zynq-7000 FPGA SoC eFuse. These options can include performing a full boot image (up to 128 KB) and performing an integrity check using the **Cyclic Redundancy Check (CRC)** algorithm. This authentication ensures that the boot image is as expected and hasn't been corrupted or tampered with. This boot time CRC computation increases the boot time by around 25 ms when using the default boot settings. Once the integrity check is performed, the BootROM code reads the boot mode setting indicated by the bootstrap pins. Then, the BootROM code reads the BootROM header from the specified external memory.

When RSA authentication is enabled, the BootROM code reads the boot image header, the Register Initialization values, and the FSBL image into the lower portion of the OCM memory (192 KB). Then, the BootROM loads the authentication public key from the boot image and then validates it by computing a **Secure Hash Algorithm (SHA)** SHA-256 signature and comparing it to the hash value stored locally in the eFuse. When the computed signature and the eFuse stored value match, the BootROM continues to the next step, which is calculating the signature for the FSBL and its authentication with the public key. However, if the computed public key signature and the value stored locally in the eFuse don't match, or the FSBL authentication fails, then the BootROM code performs a fallback to search for an alternative FSBL when the specified boot device is a NAND, NOR, or QSPI Flash. If this fallback also fails or the boot device is an SD card and the boot image was not encrypted, then the BootROM code enters an error state and enables JTAG; otherwise, it enters a secure lockdown. When the FSBL authentication is successful, the BootROM code proceeds further in the boot process.

The Zynq-7000 SoC FPGA secure configuration only supports master secure boot mode, which decrypts and authenticates the PL bitstream image and the PS firmware images using the PL AES and HMAC hard macro engines. When the RSA authentication is enabled as a boot option, the BootROM code authenticates the encrypted FSBL image with the public key stored in the eFuse before it's decrypted.

Secure booting of the Zynq-7000 SoC only allows NAND, NOR, QSPI, or SDIO as an external boot device; booting from JTAG or another external interface is not allowed in secure boot mode.

The secure boot image has the following format:

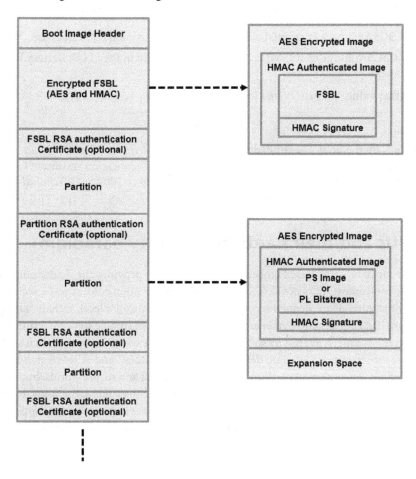

Figure 11.3 – Zynq-7000 SoC secure boot image format

The secure boot image has the following sections:

- A boot image header, which is always required for a secure boot image.

- An FSBL partition, which is always required for a secure boot image.

- An FSBL RSA authentication certificate, which is optional for a secure boot image.

- Many consecutive partitions. These host the system boot images, such as the PS firmware executable files and the FPGA bitstream files. These are optional for a secure boot image.

The boot image header word at offset 0x28 indicates whether the boot image is secure or non-secure. The value that's stored at this word location provides the source of the AES key, as follows:

- **0xA5C3C5A3**: Image encrypted using the AES key stored in the FPGA eFuse
- **0x3A5C3C5A**: Image encrypted using the AES key stored in the FPGA **Battery Backed RAM (BBRAM)**
- **Any other value**: Non-encrypted image indicating a non-secure boot

> **Information**
>
> More details on the boot image header are provided in *Section 6.3.2* of the *Zynq-7000 SoC Technical Reference Manual*. Further details on the RSA authentication certificate format and the device secure boot are available in *Chapter 32* of the *Zynq-7000 SoC Technical Reference Manual*: `https://docs.xilinx.com/v/u/en-US/ug585-Zynq-7000-TRM`.

ARM CPUs and their hardware security paradigm

To address security in ARM-based embedded systems, ARM provides the TrustZone framework, which defines two contexts of system execution – a normal execution context and a secure execution context. They make use of pre-defined security features at the hardware level, including the processor, the system peripherals, and the system interconnect. There are also specific software features that define the execution environment, as seen by software with multiple execution states known as the normal world and the secure world. These are defined in both the ARMv7 and ARMv8 architectures. The framework defines the separation methods between the normal and secure worlds, how the policy is enforced in the system architecture, and how moving between the worlds at the software level should happen. Separate software components can be running in parallel (with protected context switching) on the same hardware platform, which has both secure and non-secure components.

In this section, we will focus on the hardware aspects defined by the ARM TrustZone framework. We will look at the system and software aspects in the next section of this chapter.

ARM TrustZone hardware features

The system-wide implementation of the ARM TrustZone architecture is defined by the AMBA specifications. In most of the interconnect protocols forming the AMBA specification, such as AXI, APB, and AHB, control bits for the read and write channels are used, whereby the master requesting access to a given slave needs to set them to gain access to a peripheral defined as secure. Slave peripherals that are defined as secure will only grant access to the requesting master if these security control bits are set by the requesting read or write channel. These bits are known as **non-secure (NS)** bits, and they are defined by the AXI protocol as follows:

- **ARPROT[2:0]**: AXI Read channel protection signals
- **AWPROT[2:0]**: AXI Write channel protection signals

These signals are defined as follows:

Field	Protection level	Definition
AxPROT[0]	0b1: Privileged access 0b0: Normal access	Processor execution level or privilege level access
AxPROT[1]	0b1: Non-secure access 0b0: Secure access	An AXI master might support the security operating states and extend this concept of security to the system access. AxPROT[1] qualifies access as secure or non-secure.
AxPROT[2]	0b1: Instruction access 0b0: Data access	Specifies whether the transaction is for the data or instruction space

Table 11.1 – AXI protocol protection signals

At runtime, a requesting master sets the appropriate value to these signals, and the interconnect or the targeted slave peripheral logic decodes them and responds according to the expected TrustZone security requirements. All NS masters in the SoC must set their AxPROT[1] to 0b1, which ensures they can't access the SoC Secure slaves.

When an NS master tries to gain access to an SoC Secure slave, the system implementation is responsible for defining the type of response from the slave – that is, whether the transaction fails silently or whether it should generate an error response. An error response (*AXI SLVERR*) can be generated by the targeted Secure slave or the system interconnect port (*AXI DECERR*); this is also implementation-dependent.

The APB protocol (in the AMBA4 specifications and upward) defines the NS bits as follows using the *PPROT[2:0]* signals:

Field	Protection level	Definition
PPROT[0]	0b1: Privileged access 0b0: Normal access	Processor execution-level or privilege-level access
PPROT[1]	0b1: Non-secure access 0b0: Secure access	PPROT[1] qualifies access as secure or non-secure.
PPROT[2]	0b1: Instruction access 0b0: Data access	Specifies whether the transaction is for the data or instruction space

Table 11.2 – APB4 protocol protection signals

Like the *AxPROT[1]* signals, a requesting master sets the appropriate value for these signals, and the interconnect or the targeted slave peripheral logic decodes it and responds according to the expected TrustZone security requirements.

When an NS master tries to gain access to an SoC APB Secure slave, the system implementation is responsible for defining the type of response from the slave – that is, whether the transaction fails silently or whether it should generate an error response.

As for the AHB protocol (in the AMBA5 specifications and upwards), secure transfers have been added using an extension signal called HNONSEC, which is asserted for NS data transfers and deasserted for Secure data transfers. The HNONSEC signal is used in the address phase with the same validity constraints as the HADDR signal:

Signal	Protection level	Definition
HNONSEC	0b1: Non-secure access 0b0: Secure access	HNONSEC qualifies access as secure or non-secure.

Table 11.3 – AHB5 protocol protection signal

Like the *AxPROT[1]* and *PPROT[1]* signals for AXI and APB4, respectively, a requesting master sets the appropriate value to the HNONSEC signal, and the interconnect or the targeted slave peripheral logic decodes them and responds according to the expected TrustZone security requirements.

When an NS master tries to gain access to an SoC AHB5 Secure slave, the system implementation is responsible for defining the type of response from the slave – that is, whether the transaction fails silently or whether it should generate an error response.

The following diagram shows the usage of *AxPROT[1]* and *PPROT[1]* in a typical SoC implementation:

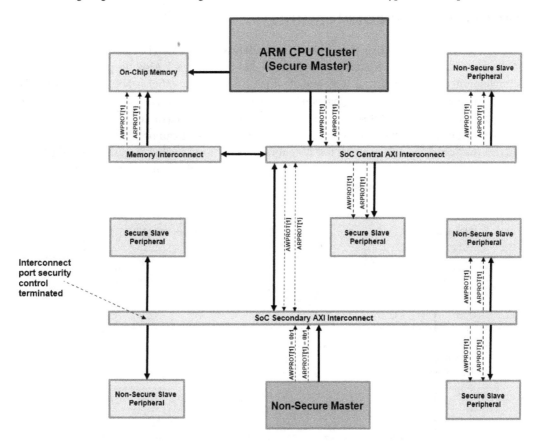

Figure 11.4 – AMBA security control bits distribution in a Secure SoC

The SoC interconnect distributes the AMBA Security bits among the SoC peripherals as it maps access to them from the different AXI masters in the SoC. Conversion of the AXI transaction signaling into the APB transaction signaling in both the request and response directions is also assured by the AXI interconnect of the SoC. In the preceding diagram, there are also slave peripherals that don't implement the AMBA security bits. However, the TrustZone architecture allows secure transaction control to be terminated and assured by the AXI interconnect ports. These interconnect port acts as a Security proxy for this type of slave peripherals without breaking the TrustZone Security protection.

Software security aspects and how they integrate the hardware's available features

From the processor perspective, the 32-bit physical address space (4 GB) is the same, but it has a 33rd bit that permits qualifying the address space accessed as Secure or NS. Also, the processor hardware is implemented as a dual virtual core, one of which is Secure; the other is NS. The hardware also implements a context-switching mechanism between the two virtual cores. This mechanism is known as **Monitor mode**. The NS bit set on the processor bus interface is a direct reflection of the actual virtual core performing the transaction. The NS virtual processor can only access NS peripherals, whereas the Secure one can view the full system resources. The following diagram shows the logical transition path between the *Normal* software execution mode and the *Secure* software execution mode, otherwise known as **Normal world** and **Secure world**, respectively:

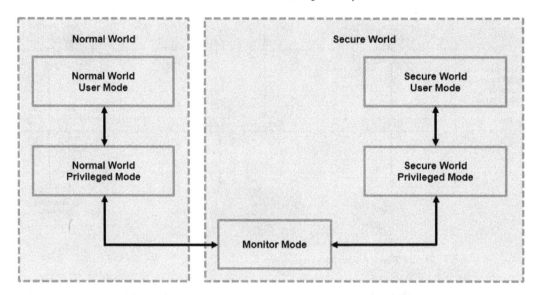

Figure 11.5 – Execution modes in the ARM processor implementing TrustZone Security

The two virtual processors can execute software on a time-sliced basis with context switching through monitor mode. Entering Monitor mode from the Normal world can only happen through exceptions. These can be triggered by executing a dedicated instruction: the **Secure Monitor Call** (**SMC**) instruction. This can also be the result of a hardware exception such as the IRQ, FIQ, External Data Abort, and External Prefetch Abort exceptions when they are configured to switch the processor to Monitor mode.

The software executing within Monitor mode is defined by the system implementation, and it is generally required to save the context of the currently executing world and restore the context of the target world to which it is switching the execution state of the processor. Branching into the target world is performed through the return-from-exception that caused entry into the target world.

The world under which the processor is executing is reflected by the NS bit of the **Secure Configuration Register (SCR)**.

Implementing a secure software architecture resides between two extremes in the possibility spectrum of the software stack running on an SoC based on the TrustZone framework. The SoC can implement a dedicated Secure world operating system, or it can simply implement a synchronous library of software code running in the Secure world. The actual implementation can be anywhere between these ends.

> **Information**
>
> This chapter briefly introduced the ARM TrustZone framework and its high-level features. To learn more about its architecture in terms of its hardware, software, and system aspects, you are encouraged to study the ARM Security Technology document at `https://developer.arm.com/documentation/PRD29-GENC-009492/c/TrustZone-Hardware-Architecture`.

Building a secure FPGA-based SoC

As already introduced, the Zynq-7000 SoC FPGA adopts the ARM TrustZone framework and provides a secure boot mechanism with a root of trust using the BootROM. It can store its public encryption and authentication keys in the eFuse provided by the FPGA, as well as use the AES and HMAC hardware engines available within the PL to be used by the PS as hard macros before the FPGA logic is even configured. The PS can securely communicate with these hard macros through the PCAP interface to accelerate the boot time process. Through the PCAP interface, the PS can decrypt, authenticate, and load the FSBL and the FPGA bitstream. These protected images are stored externally. Then, they are loaded, decrypted, and authenticated by the PS through the PCAP and then stored within the PS OCM memory to be used by the FSBL to configure the FPGA logic and continue loading the necessary firmware images needed by the SoC software. The Xilinx AXI IP peripherals also adopt the TrustZone framework by the Secure and NS access controls, which are proxied through the AXI interconnect port interfacing to the IP. This mechanism extends the Secure and NS accesses externally of the PS to the hardware accelerators built within the PL domain. The following diagram illustrates the ARM TrustZone implementation in the Xilinx Zynq-7000 SoC FPGAs:

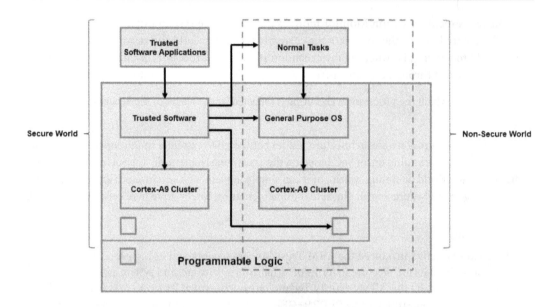

Figure 11.6 – Trusted and non-trusted software applications on a Zynq-7000 SoC FPGA

Using all the available security features based on the ARM TrustZone and the Zynq-7000 SoC FPGA specifics, we can make an SoC as secure as possible, such as the ETS SoC system we architected and built in *Part 2* of this book. The ETS SoC can greatly benefit from these security features since it is a financial system we would like to protect from external intrusion. The conceptual view of a secure Zynq-7000 FPGA-based SoC is shown in the following diagram:

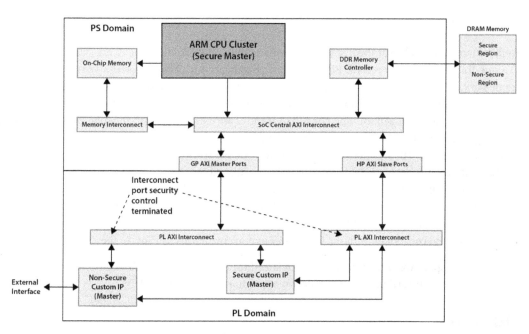

Figure 11.7 – Typical Secure Zynq-7000 FPGA-based SoC

In the preceding conceptual diagram of a Secure SoC targeting the Zynq-7000 SoC FPGA, by using its TrustZone features, Secure data can be protected from unwanted access. This can be achieved by splitting the custom IPs built within the PL domain into two categories – Secure IPs and NS IPs. The Secure custom IPs perform Secure transactions for both read and write operations from the external DRAM memory Secure region connected to the PS domain. These transactions flow through the **High-Performance AXI (HP AXI)** slave interfaces. The Cortex-A9 can also configure the PL Secure registers through the **General-Purpose AXI (GP AXI)** master interface.

The NS custom IPs are used to perform read and write operations from/to the PS DRAM memory by the external world via the HP slave world, but only to the DRAM NS region. These boundaries are configured by the Cortex-A9 via the GP master port.

When first powered up and upon PoR, the Cortex-A9 processor, following a Secure boot process, configures the custom IPs as required – that is, one as NS and the other as Secure. It also sets the registers controlling the DRAM memory regions to mark one as Secure and the other as NS. It uses the Secure DRAM region as a store for the Secure data and processor code. Once the system is running, when the NS custom IP receives data from the external world, it stores it in the DRAM NS region and notifies the Cortex-A9 processor via an inter-process communication IPC interrupt. Then, the Cortex-A9 processor sends a notification to the Secure custom IP to execute the necessary computations on the received data. Once these operations have been performed, the Secure custom IP notifies the Cortex-A9 processor of this via another IPC interrupt, which results in the Cortex-A9 sending a further notification to the NS custom IP stating that the computation is finished and that it can read the results from the NS DRAM-allocated region.

Summary

In this chapter, we looked at the key security features that are available in the Zynq-7000 SoC FPGA. We saw that these are threefold: FPGA-specific hardware features, the ARM TrustZone-based framework, and the Secure software execution environment. We covered the secure booting process and how the software and hardware images can be protected using encryption and authentication. Then, we examined how, at boot time, the FPGA features are used to establish the root of trust, decrypt the images, authenticate them, and then load the FSBL and optionally configure the FPGA with the secure bitstream file. We also delved into the Secure and NS software execution worlds and how they interact with each other via a Secure monitor. We examined the ARM TrustZone and the hardware protection mechanism used to make peripherals Secure or NS and how these transactions are qualified at runtime by the AXI and APB bus protocols. We presented a typical Secure system that combines the PS and PL domains of the Zynq-7000 SoC FPGA and how NS and Secure custom IP masters can share the data and collaborate with the Cortex-A9 to implement a Secure SoC in the Zynq-7000 FPGA.

In the next chapter, we will explore more advanced topics related to the FPGA-based SoCs with a focus on system software and its design flow when using an embedded RTOS.

Questions

Answer the following questions to test your knowledge of this chapter:

1. Describe the TrustZone framework.

2. List the Zynq-7000 SoC FPGA security-related hardware features.

3. Where does secure booting start from?

4. What is an FSBL? How can it be stored externally in a secure way?

5. Describe the main steps of securely booting a Zynq-7000-based SoC.

6. What specific control signals are used to implement security in the AMBA specifications?

7. How is security implemented in the Cortex-A9 processor as far as switching between the Secure execution environment and the NS execution environment?

8. What is a Secure monitor?

9. How is the ARM TrustZone hardware protection mechanism implemented in the PL domain?

10. How could we make the ETS SoC example design in *Part 2* of this book more secure?

12

Building a Complex Software with an Embedded Operating System Flow

In this chapter, you will learn about embedded operating system flows and discover the tools used to build a complex software application to run on an FPGA SoC. You will use the design tools available to create the SoC **board support package** (**BSP**) for the desired embedded operating system – in this case, FreeRTOS. You will go through the process of generating the SoC bootloader, which runs when the SoC is powered up and launches the embedded software. This chapter is also hands-on, so you will be guided through every step of the design process. Here, you will go through all the embedded software development phases, starting from the initial concept, followed by the actual software building, and then running it on a hardware board or a virtual platform.

In this chapter, we're going to cover the following topics:

- Embedded OS software design flow for Xilinx FPGA-based SoCs

- Customizing and generating the BSP and the bootloader for FreeRTOS

- Building a user application and running it on the target

Technical requirements

The GitHub repo for this title can be found here: https://github.com/PacktPublishing/Architecting-and-Building-High-Speed-SoCs.

Code in Action videos for this chapter: http://bit.ly/3WHLVJd.

Embedded OS software design flow for Xilinx FPGA-based SoCs

In *Chapter 8, FPGA SoC Software Design Flow*, we covered the design flow of a piece of bare-metal embedded software targeting the **Electronic Trading System** (**ETS**) SoC. In this chapter, we will go over the steps required to build a **Real-Time Operating System** (**RTOS**)-based software application targeting an FPGA SoC. We will perform the entire design flow within the Vitis environment and choose FreeRTOS as the embedded **operating system** (**OS**). The design flow can start from a hardware design previously performed in the Vivado environment that has been exported as an XSA archive file (such as the ETS SoC design example). This is then chosen in the Vitis IDE as hosting hardware for the SoC. An alternative method is to choose a Zynq-7000 SoC demo board as the target hardware and perform the necessary steps to create the required framework, as will be detailed in this chapter. Follow these steps to design a FreeRTOS-based SoC software:

1. Launch the Vitis IDE from within the Linux Virtual Machine environment using the following command:

    ```
    $ sudo /<install path of Vitis>/Xilinx/Vitis/2022.1/bin/
    vitis
    ```

2. Once the Vitis IDE is up and running, go to **File** | **Create new platform project**. The following wizard will open. Give the new platform a name, such as RTOS_SoC:

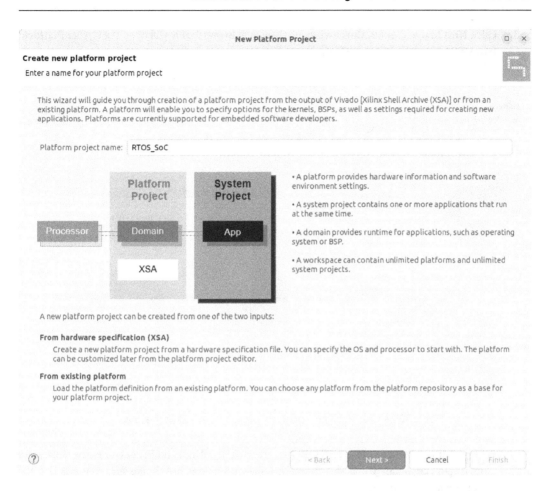

Figure 12.1 – Creating a new platform project in the Vitis IDE

3.   Click **Next** to open the **Platform** window, then click **Browse**. This will open the **Create Platform from XSA** menu. Select **zc702.xsa**, then click **Open**:

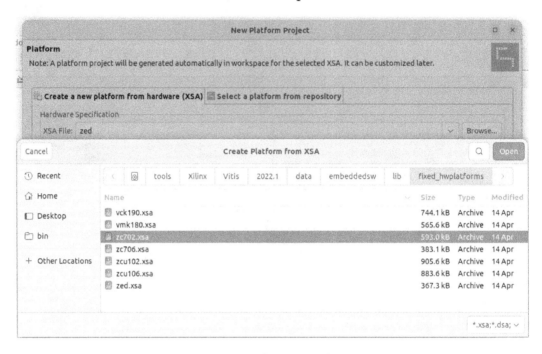

Figure 12.2 – Creating a new platform project from XSA in the Vitis IDE

4.   Set **Operating system** to the FreeRTOS port (**freertos10_xilinx**) and **Processor** to the Cortex-A9 Core 0 (**ps7_cortexa9_0**), as shown in the following screenshot. Make sure that you tick the **Generate boot components** option. Then, click **Finish**:

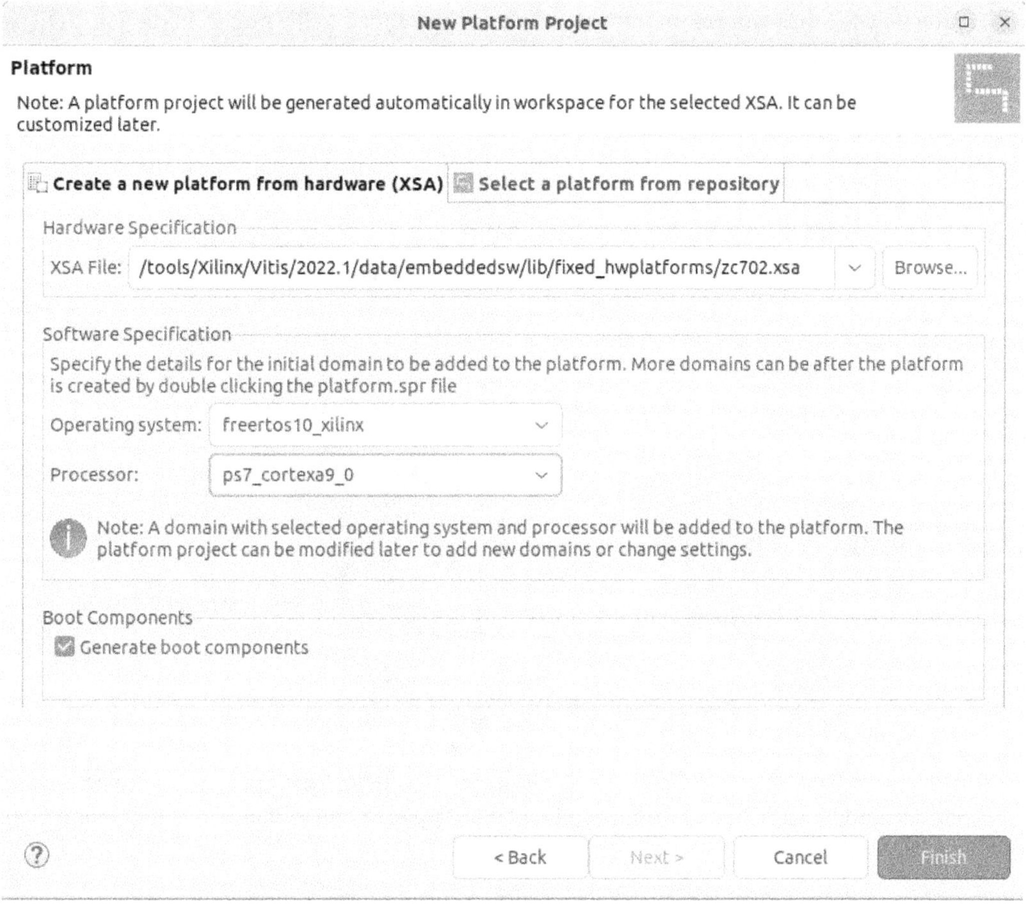

Figure 12.3 – Selecting the OS and the processor for the new platform project in the Vitis IDE

5.   The **RTOS_SoC** platform project will be created and added to the project in the Vitis IDE, as shown in the following screenshot:

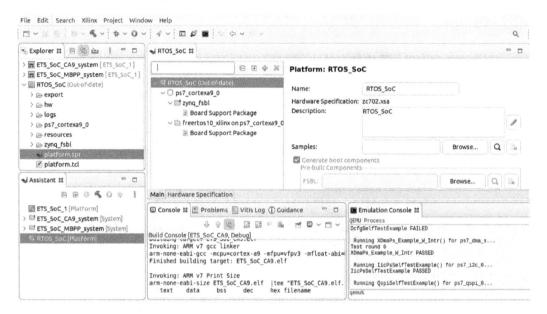

Figure 12.4 – Newly created RTOS_SoC platform project in the Vitis IDE

6.   Now, we can create an application project using the Vitis IDE's **New Application Project** wizard, which can be launched by going to **File | New Application Project**. Once it has launched, click **Next**:

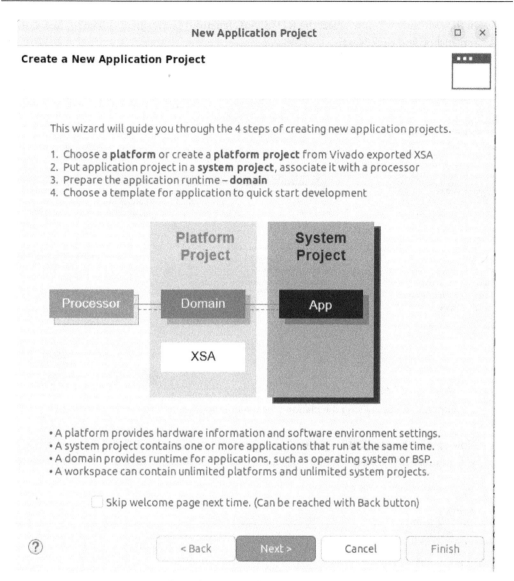

New Application Project   □ ✕

Create a New Application Project

This wizard will guide you through the 4 steps of creating new application projects.

1. Choose a **platform** or create a **platform project** from Vivado exported XSA
2. Put application project in a **system project**, associate it with a processor
3. Prepare the application runtime – **domain**
4. Choose a template for application to quick start development

Platform Project / System Project

Processor — Domain — App

XSA

• A platform provides hardware information and software environment settings.
• A system project contains one or more applications that run at the same time.
• A domain provides runtime for applications, such as operating system or BSP.
• A workspace can contain unlimited platforms and unlimited system projects.

☐ Skip welcome page next time. (Can be reached with Back button)

?   < Back   Next >   Cancel   Finish

Figure 12.5 – Launching the New Application Project wizard in the Vitis IDE

7. In the following window, select the **RTOS_SoC [custom]** platform, as shown in the following screenshot, and click **Next**:

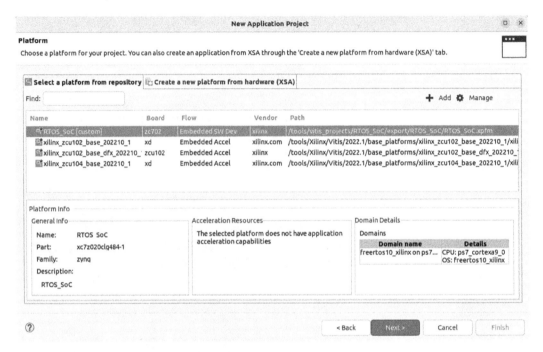

Figure 12.6 – Specifying the platform for the new application project in the Vitis IDE

8. Now, specify the **Application project name** property, as shown in the following screenshot. This will automatically assign it to the Cortex-A9 core 0, which is fine in this case. Then, click **Next**:

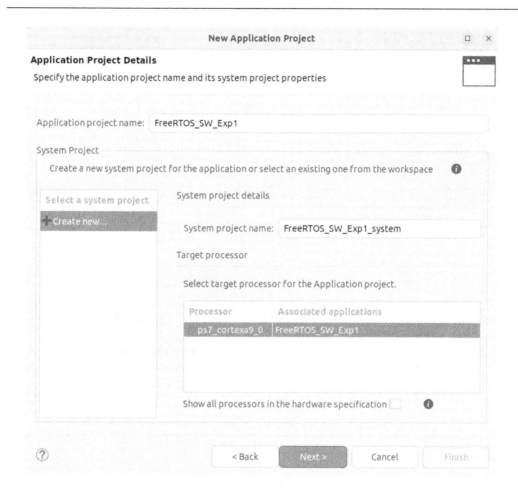

Figure 12.7 – Specifying the new application project name in the Vitis IDE

9. Next, we need to select the domain for the application project. Since we only have a single domain (**freertos10_xilinx on ps7_cortexa9_0**) in this project, leave the default values as-is and click **Next**:

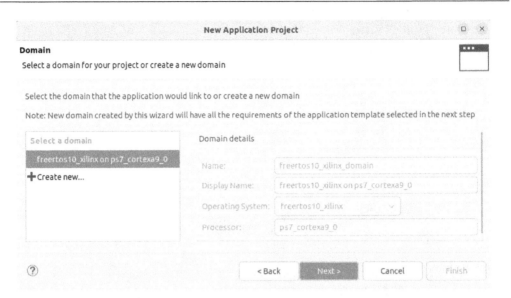

Figure 12.8 – Specifying the domain to associate with the new application project

10. Now, you can choose a template application project to start with. Here, select the **FreeRTOS Hello World** application since this is a good learning example to gain familiarity with simple multi-tasking and message passing between tasks. Then, click **Finish**:

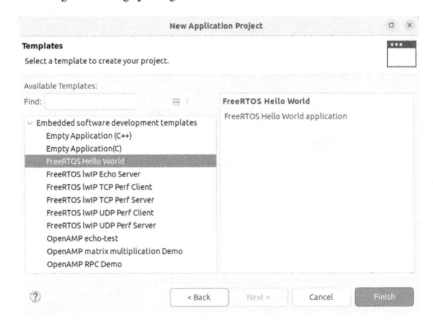

Figure 12.9 – Selecting the FreeRTOS Hello World template for the application project

11. The FreeRTOS application project, including the design template, will be added to the **Vitis Explorer** list, as shown here:

Figure 12.10 – FreeRTOS application project added in Vitis Explorer

With that, we have generated the BSP, the bootloader, and the FreeRTOS application projects. Next, we will delve into the BSP and the bootloader, before examining the application software building process in the last section of this chapter.

Customizing and generating the BSP and the bootloader for FreeRTOS

It is important to understand the layout of an RTOS-based software implementation of an FPGA-based SoC and its different components within the Vitis environment. We know from the previous chapter that on **Power-on-Reset (PoR)** or following a global system reset, the Cortex-A9 core 0 processor boots from the BootROM in a secure way. Depending on the security settings, a secure boot is started if instructed to do so; if not, a **Non-Secure (NS)** boot takes place. However, under whichever specified boot mode (secure or NS) by **eFuse** or the **Battery-Backed RAM (BBRAM)**, the **First Stage Boot Loader (FSBL)** is the next image to be loaded by the BootROM from **Non-Volatile Memory (NVM)** NAND, NOR, or QSPI Flash. We also know that an FPGA bitstream can optionally be loaded from the NVM and prepared for configuring the FPGA logic according to the security settings. Finally, it is the FSBL that decides what other images to load from NVM and how to prepare them according to the system security settings. These images include the application software to run on the SoC. Also, running software on FreeRTOS requires the use of a specific BSP that Xilinx has ported using the FreeRTOS packages and its own libraries, as well as the IP drivers used in the design. The Vitis IDE automates a lot of customization under the hood and lets you choose the components you want services from, as well as the minimum required to boot FreeRTOS and run its services on the FPGA SoC. The following steps will show you how to create the BSP for our FreeRTOS application project, as well as provide an insight into its visual structure and included components that we can customize if needed by our end application:

1. Recall that in *Step 4* of the previous section, we ticked the **Generate boot components** option, as shown in *Figure 12.3*. This selection is crucial to create the necessary boot applications and associated components to be able to start the Cortex-A9 and the SoC properly to run FreeRTOS after the PoR or a global system reset. Once the **RTOS_SoC** platform project has been created and added to the project in the Vitis IDE, the boot project (**zynq_fsbl**) is created, including its BSP, as shown in the following screenshot. We have also created the BSP for the FreeRTOS application project.

Figure 12.11 – BSPs for the FSBL and FreeRTOS application projects

2. As we explored in *Chapter 8, FPGA SoC Software Design Flow*, for bare-metal applications, we can explore the FreeRTOS BSP components and customize them when necessary. The following screenshot shows the BSP and the Vitis interface to modify it:

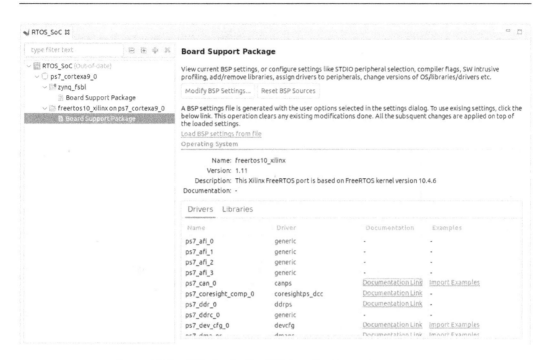

Figure 12.12 – BSPs for the FreeRTOS application project summary

3. Click **Modify BSP Settings…** to enter the Vitis FreeRTOS BSP customization interface. The **Board Support Package Settings** window will open. Explore the different components and leave everything as-is by default in Vitis for the FreeRTOS **Hello World** application example:

Figure 12.13 – The Board Support Package Settings window

4.   Now, we can generate the BSPs for both the bootloader and the FreeRTOS application projects. In the Vitis IDE's **Explorer** window, right-click the **RTOS_SoC** domain entry and click **Build project**, as shown in the following screenshot:

Figure 12.14 – Building the BSP components for FSBL and FreeRTOS

This will generate the necessary FSBL components and libraries required by the application building process for FreeRTOS.

Building a user application and running it on the target

Now, we will build the FreeRTOS *Hello World* application and run it on the virtual platform within the Vitis IDE. By doing so, we can revisit the application template source code and build upon it so that it matches our software requirements. For the ETS SoC project, we can generate another FreeRTOS application based on a UDP client template provided by the Vitis project creation wizard and use it as a starting example. It is by itself a proper software project that requires modifying the Xilinx Ethernet drivers so that they match the microarchitecture requirements of our design. We have access to the driver's source code and the UDP client template to achieve this. As an exercise for any embedded software developers using this book, it can be a nice challenge to implement the software design presented in *Chapter 8, , FPGA SoC Software Design Flow,* by using the knowledge you've accumulated thus far. From the **Electronic Trading Market (ETM)** side, the **User Datagram Protocol (UDP)** server example can be used as a starting template, though any other UDP server software from the open source community can be implemented on a Linux host to act as the ETM and provide the UDP streams as per the defined ETM protocol.

To build the FreeRTOS software application in the Vitis IDE and run it on the virtual platform, follow these steps:

1. Right-click the **FreeRTOS_SW_Exp1** project entry under **FreeRTOS_SW_Exp1_system** and click **Build**. We will use the default project settings set by the Vitis IDE for this template software application project:

Figure 12.15 – Building the FreeRTOS project application

2. The compiler will run all the necessary steps and build all the required object files to generate the FreeRTOS software executable image, as shown in the Vitis **Console** window. The application executable image is around 171 KB:

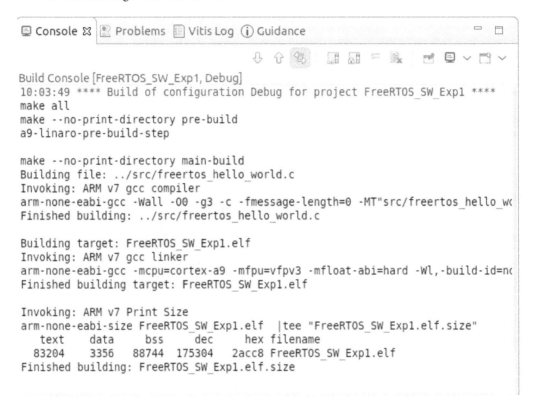

```
🖥 Console ⊠  🗐 Problems  🗐 Vitis Log  ⓘ Guidance            ⊟  🗖

                            ⬇ ⬆ 🔄    🗔 🗔 = 🗎   🖳 🖳 ⌄ 🗎 ⌄

Build Console [FreeRTOS_SW_Exp1, Debug]
10:03:49 **** Build of configuration Debug for project FreeRTOS_SW_Exp1 ****
make all
make --no-print-directory pre-build
a9-linaro-pre-build-step

make --no-print-directory main-build
Building file: ../src/freertos_hello_world.c
Invoking: ARM v7 gcc compiler
arm-none-eabi-gcc -Wall -O0 -g3 -c -fmessage-length=0 -MT"src/freertos_hello_wc
Finished building: ../src/freertos_hello_world.c

Building target: FreeRTOS_SW_Exp1.elf
Invoking: ARM v7 gcc linker
arm-none-eabi-gcc -mcpu=cortex-a9 -mfpu=vfpv3 -mfloat-abi=hard -Wl,-build-id=nc
Finished building target: FreeRTOS_SW_Exp1.elf

Invoking: ARM v7 Print Size
arm-none-eabi-size FreeRTOS_SW_Exp1.elf  |tee "FreeRTOS_SW_Exp1.elf.size"
   text    data    bss    dec     hex filename
  83204    3356  88744 175304   2acc8 FreeRTOS_SW_Exp1.elf
Finished building: FreeRTOS_SW_Exp1.elf.size
```

Figure 12.16 – Vitis IDE console showing the FreeRTOS software image being built

3. Let's launch the virtual platform emulator and run the FreeRTOS application software we have just built for it. Right-click the **FreeRTOS_SW_Exp1** project entry, select **Run As**, and then select **2 Launch SW Emulator (Single Application Debug)**:

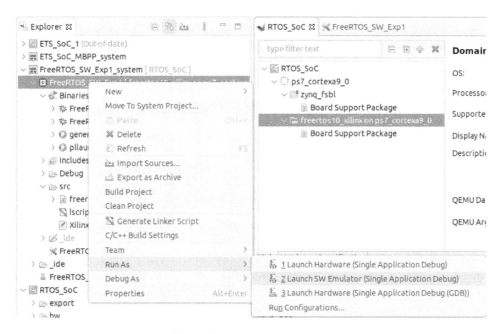

Figure 12.17 – Launching the FreeRTOS application image on the virtual platform in the Vitis IDE

4.   This will automatically launch the QEMU-based virtual platform in the Vitis IDE, boot the Zynq-7000 SoC, and load the FreeRTOS software application, as shown here:

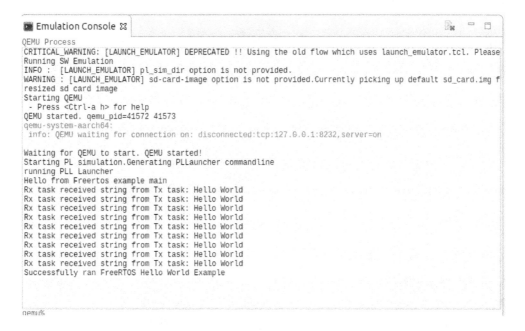

Figure 12.18 – Running the FreeRTOS application image on the virtual platform in the Vitis IDE

The FreeRTOS *Hello World* application will run on the virtual platform to completion, as shown in the preceding screenshot.

This will complete the complex software design flow in the Vitis IDE, from the initial platform specification to running the FreeRTOS application generated as a template by the Vitis IDE. In this example, we used the Vitis-included QEMU virtual platform, which will help greatly in learning the details of developing an RTOS-based embedded software for advanced high-speed SoCs targeting the Xilinx Zynq-7000 SoC FPGAs.

Summary

In this chapter, we looked at the key steps required to build a complex RTOS-based software application and specifically its associated software design flow within the Vitis IDE. We covered all the steps required in this process. We also looked at the platform generation using an existing XSA file for a known demo board to generate a platform and its associated project domain. Then, we learned how to generate a software application associated with the created domain and showed the necessary settings for the FreeRTOS embedded OS. We also generated an application software project example to run on FreeRTOS. We delved into the bootloader topic and how to create and customize the BSP for both the FSBL and FreeRTOS software application projects. Then, we built and ran these software projects on the QEMU virtual platform within the Vitis IDE. The output loggings from the software application on the QEMU console echoed a successful software application run.

In the next chapter, we will explore more topics related to FPGA-based SoC advanced applications with a focus on digital signal, image, and video processing.

Questions

Answer the following questions to test your knowledge of this chapter:

1. What is an RTOS?
2. What should be done to specify the ETS SoC hardware design as the target platform for FreeRTOS-based software projects?
3. What is a BSP? List its components.
4. How can we generate the boot components for the FreeRTOS project in the Vitis IDE?
5. List the required steps for generating a LwIP Perf UDP client FreeRTOS-based software application.
6. List the steps for customizing the BSP.
7. What is an FSBL? When is it involved?
8. Describe the steps performed by the FSBL we generated for the RTOS_SoC platform.
9. List the steps you must follow to build the FreeRTOS software example.
10. How can we target the QEMU virtual platform to run FreeRTOS-based software projects in the Vitis IDE?

13

Video, Image, and DSP Processing Principles in an FPGA and SoCs

In this chapter, you will learn about some of the advanced applications implemented in modern FPGAs and SoCs and what makes these devices such powerful compute engines for these types of processing- and bandwidth-demanding applications. You will gain clarity on how the parallel processing required by **Digital Signal Processing (DSP)** applications in general can be easily implemented in the FPGA logic and how these parallel compute engines can be interfaced to wide memories and the powerful CPUs available in the SoCs. This chapter is purely informative and introduces high-level architectural details that may inspire you in designing and building these kinds of applications.

In this chapter, we're going to cover the following main topics:

- DSP techniques using FPGAs

- DSP in an SoC and HW acceleration mechanisms

- Video and image processing implementation in FPGA devices and SoCs

DSP techniques using FPGAs

Performing DSP in an FPGA or an SoC-based FPGA, such as the Zynq-7000 SoC or the UltraScale+ MPSoC, is no different than performing it in an **Application-Specific Integrated Circuit (ASIC)** specifically built for such operations. There are many advantages to performing DSP operations in an FPGA technology in comparison to using an ASIC and these are mostly due to the flexibility, extensibility, and scaling advantages of using an FPGA-based DSP solution. Classically, FPGAs were chosen to implement DSP computation units for many industries, telecommunication being one of the dominant ones. Wireless communication standards were still evolving and the time to market was, and is still, an important business objective, making using a flexible solution, such as FPGA-based

DSP, an attractive option. In these applications, most of the time, the FPGA device was a companion chip to a powerful processor. The architecture of these solutions evolved around reconfigurability as software running on the processor could be changed and upgraded even post the product deployment, and so are the FPGA DSP compute engines changeable and upgradeable. These types of architectures are still in deployment, but some users sometimes have preferences for using a processor architecture that is not based on ARM; however, these users are in the minority nowadays. The following diagram helps in visualizing the simple concept of a dual-chip solution to solve most of the DSP-intensive design challenges:

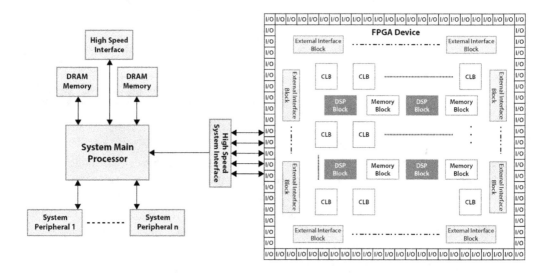

Figure 13.1 – Conceptual diagram of an electronic system using a dual-chip solution for DSP

Most applications requiring a processor and a companion FPGA to perform hardware acceleration and specifically intensive DSP computations will find an FPGA-based SoC attractive for performing their DSP operations. As we already know, these types of devices combine both a powerful processor and a rich logic elements resource. This combination and integration of both the processor subsystem and the FPGA logic for the DSP implementation within the same device has many advantages, such as the following:

- Reduction of the **Bill of Materials (BOM)** requirements
- Optimization of the overall solution's power consumption and power design
- Enhancement of the performance in terms of throughput between the two domains
- Reduction of the electronics board size and thus the overall product volume

Zynq-7000 SoC FPGA Cortex-A9 processor cluster DSP capabilities

FPGA-based SoCs, such as the Zynq-7000 SoC, offer a powerful DSP operations-capable engine within the Cortex-A9 cluster called NEON **Media Processing Engine (MPE)**. NEON designates an ARM **Intellectual Property (IP)** used for the Advanced **Single Instruction Multiple Data (SIMD)** functionality of ARMv7 and ARMv8 Cortex-A and Cortex-R processors. SIMD is a generic term referring to performing in parallel multiple operations on similar datasets when the dataset is of a smaller size than the processor width. A simple example is to perform four parallel addition operations (ADD4) on 8-bit datasets using a 32-bit processor. In this example, the single instruction is the ADD4 operation, and the multiple datasets would be the two 32-bit vectors holding each of the four 8-bit datasets. The following figure illustrates the basic SIMD concept on a 32-bit processor:

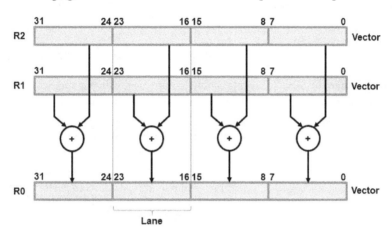

Figure 13.2 – SIMD example for a four-way 8-bit unsigned integer ADD operation

In the ARMv7 architecture specification, the Advanced SIMD extension was added to the ARMv7-A and ARMv7-R profiles. The SIMD concept is further expanded by the addition of new instructions, which take as operands 64-bit vectors (doubleword) and 128-bit vectors (quadword). These Advanced SIMD instructions operate on 64-bit and 128-bit registers and their implementation by ARM is called **NEON**. The following figure shows the NEON ADD16 operation:

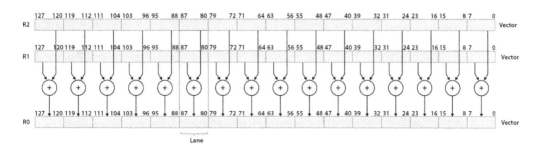

Figure 13.3 – NEON example for a 16-way 8-bit unsigned integer ADD operation

The preceding simple example is provided to visualize the principles of the NEON implementation in the ARM architecture. This helps in understanding the basis of performing the fundamental mathematical processing using a 32-bit processor but operating on wider vectors of datasets in parallel. These provide a huge performance improvement in comparison to sequentially performing the same operation on a pair of datasets at a time; however, the software code and the data representation should be optimized or vectorized to map well to NEON use.

> **Information**
>
> The following article provides more information on NEON architecture, implemented operations, and the supported data types: `https://developer.arm.com/documentation/dht0002/a/Introducing-NEON/What-is-SIMD-/ARM-SIMD-instructions`.

Zynq-7000 SoC FPGA logic resources and DSP improvement

Exploiting parallelism in the SIMD concept is going to improve the throughput when there is a chance to perform the same operation on multiple datasets. However, there is no possibility to pipeline many of these operations and avoid storing back to memory and loading from it. Data movements from/to the system memory are required to perform successive operations when using a processor, even with a SIMD engine, whereas by using an FPGA-based SoC, this kind of multiple SIMD operations pipelining is easy to implement. The following figure illustrates the extra capability of pipelining DSP operations and avoiding storing back to memory and reloading datasets to perform the next series of operations. This is also possible as we can extend the computation units to any number required to implement the desired parallel and pipelined DSP architecture.

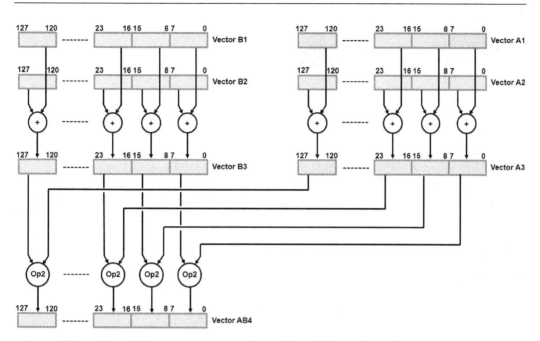

Figure 13.4 – DSP computation using the FPGA parallelism and pipelining capabilities

The FPGA logic elements, DSP slices, and distributed registers provide an extensive pool of resources to implement any mathematical function in a cost-effective way that is hard to beat using a classical processor, even with its attached coprocessing units, such as SIMD engines. When using a Zynq-7000 SoC FPGA, the computation algorithm can be mapped in such a way that sequential, control, and house-keeping operations are run on the Cortex-A9, whereas intensive mathematical operations are performed by the FPGA logic. The FPGA logic can implement wide and deep DSP engines that can operate on a large bank of datasets.

Zynq-7000 SoC FPGA DSP slices

The Zynq-7000 SoC FPGAs include many DSP slices, depending on the device size and density. The DSP slice architecture looks as in the following figure:

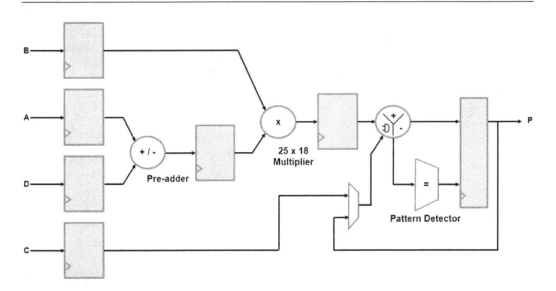

Figure 13.5 – Zynq-7000 SoC FPGA DSP slice architecture

The DSP slice contains a 25 x 18-bit two's complement multiplier and 48-bit accumulator. When not needed, the multiplier can be bypassed, resulting in two 48-bit vector inputs to the SIMD engine. The SIMD arithmetic unit can operate on a dual 24-bit or quad 12-bit dataset per vector and is able to perform addition, subtraction, and accumulator operations. It is also capable of performing logic operations.

> **Information**
>
> Details of the Zynq-7000 SoC DSP slices are available at https://docs.xilinx.com/v/u/en-US/ug479_7Series_DSP48E1.

DSP in an SoC and hardware acceleration mechanisms

DSP computation extensibility using the FPGA logic resources is just a special case of the hardware acceleration techniques covered in the previous chapters of this book. It is therefore a matter of architecture design to split and coordinate what will be running on the Cortex-A9 embedded software and what computation will be shifted to the FPGA logic resources implementing the DSP engines.

Accelerating DSP computation using the FPGA logic in FPGA-based SoCs

In an FPGA-based SoC such as the Zynq-7000 SoC, DSP computation can be implemented using the FPGA logic and DSP resources. The SoC architecture should define how the shared data to operate on should be moved around the SoC, how the results shall be shared with the Cortex-A9, and any external entity that the Zynq-7000 SoC FPGA interfaces with. Obviously, it is also important to design an **Inter-Processor Communication (IPC)** mechanism that is optimal and avoids any system bottlenecks. The following diagram provides a system overview of the DSP computation extensibility and the possible data paths and communication paths in the Zynq-7000 SoC FPGA:

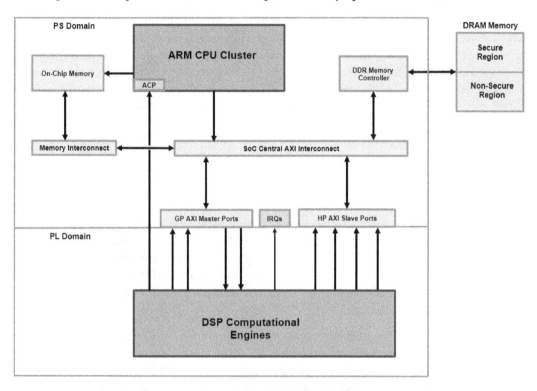

Figure 13.6 – Zynq-7000 SoC FPGA DSP computation integration

As shown by the preceding figure, and like the already-introduced hardware acceleration integration techniques in the previous chapters of this book, the DSP computation engines can be integrated into the SoC design using the same approach. The interfacing can be implemented over the **High-Performance (HP)** 64-bit AXI slave ports for high-throughput data accesses from the DSP engine masters to the SoC domain and external memory. Low- and medium-throughput operations from the DSP engines can make use of the **General-Purpose (GP)** 32-bit AXI slaves. From the **Processing Subsystem (PS)** masters, the GP 32-bit AXI masters provide a gateway to access the DSP engine address space. For a cache-coherent data exchange and sharing between the DSP computation engines and the Cortex-A9 processor, the 64-bit AXI slave ACP port can be used.

Video and image processing implementation in FPGA devices and SoCs

Video processing, specifically real-time video processing, requires intensive DSP computation. In the last decade, we started observing the proliferation of these applications in embedded systems, which possess a limited amount of computation, storage, and power resources. The emergence of IoT and distributed systems is adding to the abundance of computationally demanding devices with these limited resources. Many architectures are also evolving to solve this dilemma and balance the processing requirements and the limited resources in these devices. Several applications, such as object detection, video surveillance, machine vision, and security, are using FPGA-based SoCs where the PS implements the device security, communication, and control, whereas DSP-intensive operations are offloaded to the FPGA logic to implement the computationally intensive algorithms. This approach is helping to minimize the time to market and the overall solution cost.

Xilinx AXI Video DMA engine

Xilinx provides the AXI **Video Direct Memory Access (VDMA)** IP core, which is a **Direct Memory Access (DMA)** engine adapted for video applications. It can perform video data read and write transfer operations from an AXI mapped address space to an AXI4-Stream destination. It can also execute transfers in the opposite direction. This IP integrates features that are specific to the video systems. These include genlock and frame sync for fully synchronized frame DMA operations as well as two-dimensional DMA transfers.

Both the scatter/gather or register direct mode operations are provided by the AXI VDMA IP for data transfers controlled by the central processor. The AXI VDMA has an AXI4-Lite slave interface for the IP initialization, status, and control registers access. The AXI VDMA core can handle data conversion and padding from the common 24-bit format in video applications to the 32-bit generic AXI interface. The AXI VDMA architecture is shown in the following figure:

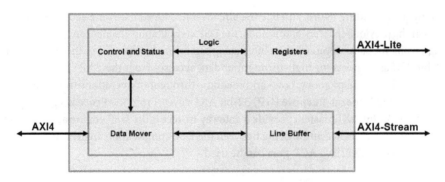

Figure 13.7 – Xilinx AXI VDMA IP architecture

> **Information**
>
> Further details on the Xilinx AXI VDMA engine are available from `https://docs.xilinx.com/v/u/6.2-English/pg020_axi_vdma`.

Video processing systems generic architecture

Video processing systems usually contain a video stream source, video processing performed by the SoC hardware and coordinated by the SoC software, and a destination. The processing is performed in multiple stages in a pipelined fashion. These stages consist of an input interface stage, a video preprocessing step, a main processing step, a postprocessing step, and an output interfacing stage. At most of these steps, common memory accesses are required at the video quality and resolution rate. The following figure illustrates the generic architecture of a digital video processing system:

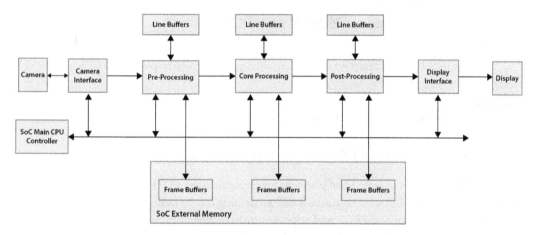

Figure 13.8 – Typical digital video processing system pipelined architecture

For further information on designing HP video systems with the Zynq-7000 SoC FGPAs, you are encouraged to study Xilinx application note XAPP1205 available at `https://docs.xilinx.com/v/u/en-US/xapp1205-high-performance-video-zynq`.

Using an SoC-based FPGA for edge detection in video applications

In video processing applications, edge detection is a technique used to extract information from the frames as a base step for feature extraction and the segmentation of objects. This process delimits the object among others and the image background. In these kinds of systems, the input is received from a CMOS camera, the processing is performed in the SoC FPGA, and the output is streamed to a display unit. The following diagram provides the architecture of an example system:

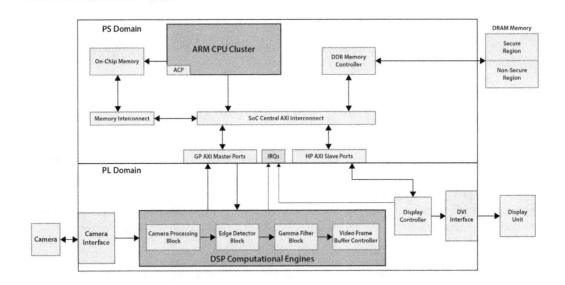

Figure 13.9 – Edge detection video application architecture example in an FPGA SoC

The preceding architecture provides a powerful platform for edge detection video applications and optimizes the overall device cost and power consumption.

Using an SoC-based FPGA for machine vision applications

The Zynq-7000 SoC FPGA is an ideal platform for implementing an HP machine vision algorithm such as the MVTec HALCON. The vision-based processing can be offloaded to the **Programmable Logic (PL)** portion of the SoC FPGA and achieves orders of magnitude of improvement of software-only-based machine vision processing. The following architecture can be implemented in the Zynq-7000 SoC FPGA:

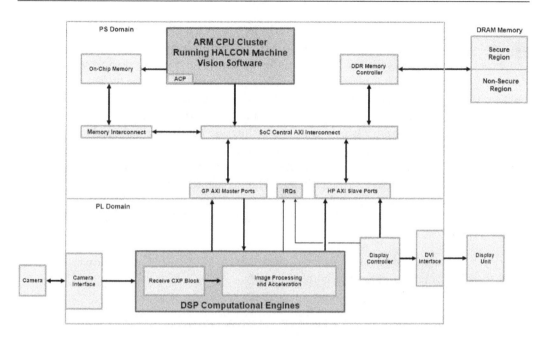

Figure 13.10 – Machine vision application architecture example in an FPGA SoC

This architecture accelerates the vision processing with visual applets implemented in the PL portion of the FPGA SoC and HALCON software running on the Cortex-A9 in the PS. It exceeded 90 frames/second in a machine vision model for an industrial application for manufactured parts inspection. This acceleration is around 20x faster in comparison to a software-only-based machine vision application.

> **Information**
>
> The preceding architecture example is provided in a Xilinx white paper. Further details on this specific video application can be found at `https://docs.xilinx.com/v/u/en-US/wp453-machine-vision`.

Summary

In this chapter, we looked at some advanced applications where FPGA-based SoCs are well suited as a single-chip architecture with a fast time to market product development and a lower cost solution, which is also lower power in comparison to multi-chip-based architectures. These advanced applications find many uses in DSP and video and image processing systems. Like the generic hardware acceleration capabilities of FPGA-based SoCs, DSP applications are well suited to these types of devices. SoCs built using PL have tight integration between the Cortex-A9 CPU cluster and the PL. This flexible architecture offers a scalable DSP solution where the exact amount of compute capabilities is used. Designers can start with a pure software solution, then they can offload heavy-compute DSP and video and image processing algorithms to the FPGA logic, which is rich in DSP building blocks. We examined the intrinsic DSP capabilities of the Cortex-A9 CPU, which make use of the Advanced SIMD feature of the ARM architecture implemented under the name NEON. These use vectors of multiple datasets upon which the same operation is simultaneously executed. We also looked at the case of multiple parallel vector operations being needed to improve the performance of the Cortex-A9 NEON engine and how this can be achieved using the DSP building blocks in the FPGA logic. We looked at a generic video and image processing system architecture and its different building blocks. We also looked at a few video and image processing applications and how they can be implemented in an FPGA SoC. We saw that many applications in the video and image processing domain are well suited for SoC FPGA implementation where heavy algorithms compute engines are built using the FPGA logic DSP resources and the software management and orchestration are running on the Cortex-A9 CPU. We then looked at two example architectures using the Zynq-7000 SoC FPGA, the first for edge detection and the second for machine vision.

In the next chapter, we will explore more topics related to the FPGA-based SoC advanced applications with a focus on communication and control systems.

Questions

Answer the following questions to test your knowledge of this chapter:

1. List the advantages of using an FPGA-based SoC to implement a system with the heavy use of mathematical algorithms in a DSP application.

2. Describe the process of deciding whether offloading the DSP computation to the FPGA logic is the correct decision to make. You can refer to the hardware acceleration methodologies described in the previous chapters.

3. Define SIMD and describe how it can improve the software runtime performance and reduce power consumption.

4. List some of the challenges you may face if you decide to use a CPU SIMD feature.

5. What does NEON refer to in the ARM architecture and how wide are the datasets its instruction set can operate on?

6. How could you extend the NEON capabilities of the Cortex-A9 CPU in a Zynq-7000 SoC FPGA? Name the main logic resources you would use to achieve this.

7. Draw a simple SoC diagram where you offload heavy DSP algorithms to the FPGA logic, showing both the data path and the control path.

8. Describe the minimal system architecture of a video processing system. Does every video processing system need to have all the listed subcomponents?

9. What is the Xilinx VDMA engine? How different it is from a central DMA engine for moving regular data in the SoC?

10. What is an edge detection video application? List its major architecture components.

11. Which components of the edge detection video processing application are well suited for the FPGA logic implementation? Can you explain why?

12. Name one of the machine vision algorithms. What kind of performance improvement should we expect when we offload its heavy compute algorithms from the Cortex-A9 CPU to a DSP engine implemented in the FPGA logic?

14

Communication and Control Systems Implementation in FPGAs and SoCs

In this chapter, you will continue learning about more advanced applications implemented in modern FPGAs and SoCs and what makes these devices suitable to host these types of processing- and bandwidth-demanding applications. You will gain clarity on how pipelined and parallel processing engines required by modern wired and wireless communications systems in general can be easily implemented in the FPGA logic. You will also examine how these parallel compute engines can be interfaced to wide memories and integrated with the powerful CPUs available in the SoCs. You will also gain an overview of modern control and industrial systems, which use a wide range of custom and industry-specific standards. This chapter is also purely informative and introduces high-level architectural details that may inspire you in designing and building these applications. This is the closing chapter of *Part 3*, which has covered most of the FPGA-based SoC advanced features and advanced applications. It is also the closing chapter of this book; hopefully, by now you have learned enough about the theoretical, architectural, and practical aspects of FPGA-based SoCs. I hope that you are now able to start architecting, designing, and implementing your next SoC-based product.

In this chapter, we're going to cover the following main topics:

- Communication protocol layers
- Communication protocol layers mapping onto FPGA-based SoCs
- Control systems overview
- Control systems hardware and software mappings onto FPGA-based SoCs

Communication protocol layers

Communication protocols used in electronics systems and the telecommunication industry often follow the **Open Systems Interconnection** (**OSI**) model. The OSI model divides the communication protocols into seven layers. These layers are stacked on top of each other and have an abstracted interface through which the communication flows from one layer to the next.

OSI model layers overview

In real implementations, some communication protocols collapse some layers together when it doesn't make sense to divide the segment of the protocol any lower than the chosen implementation. Also, a communication protocol is often a concatenation of many other protocols that collaborate to map to an OSI communication model. The OSI model layers are as follows:

- Layer 1: Physical layer
- Layer 2: Data Link layer
- Layer 3: Network layer
- Layer 4: Transport layer
- Layer 5: Session layer
- Layer 6: Presentation layer
- Layer 7: Application layer

The Physical layer is the communication medium interface. It is responsible for the packet data conversion into the medium signal format, and transferred over it in the supported format to reach and be understood by the link partner.

The Data Link layer is responsible for interfacing with the network interface. Its main role is to provide the logical links with the network.

The Network layer is a gateway via which the data flow transitions between layers by means of logical addressing and data packet routing.

The Transport layer provides the mechanisms of the data transmission. It controls the data exchange flow and manages errors in the transmission.

The Session layer maintains communication sessions between applications on the connected nodes.

The Presentation layer formats the data stream to transmit and extract data content from the received packets, encrypt and decrypt the data content when applicable, and manage other data formatting (compression and decompression) in use by the protocol.

The Application layer is the communication protocol entry point and provides connectivity options to the software applications for transferring data, control, and messages to another application running on a different node or the same one but behind the **Operating System (OS)**. The following figure illustrates the concept of the layered approach in communication protocols conforming to the OSI model:

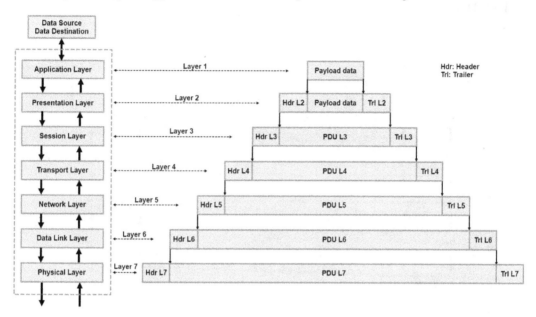

Figure 14.1 – Peer-to-peer communication using the OSI model

The Application layer receives the data to transfer to another node from a single source or multiple sources. It then encapsulates it in a format called **Protocol Data Unit (PDU)**, which defines what header and trailer are added to the data. The PDU also defines how long the payload data should be. All these are protocol specific. The communication data is passed from the Application layer through all the intermediate layers to reach the Physical layer, and at each transition, a protocol encapsulation is used when the data crosses the boundary from layer *L* to layer *L-1* in the transmit direction. For the receive direction, a symmetrical path is taken; this time, data is passed from layer *L-1* to layer *L*, and each time the information added from the protocol layer is removed. As mentioned earlier, in certain communication protocols, some layers of the OSI model are merged to simplify the protocol implementation. The Physical layer is responsible for putting data outside of the node on the medium or the link, which will then transfer it to the next node in the communication topology used.

Communication protocols topology

In communication, we can distinguish between peer-to-peer and switched communication topologies. The peer-to-peer topology is the simplest of them all and, as its name implies, it consists of directly connecting two physical nodes. The switched topology connects many nodes in a network and has many nodes and protocols in between the communicating nodes at any given time performing the associated switching operations.

In peer-to-peer communication, the communicating devices are connected directly without any intermediary device, and they exchange data directly through their physical layers over a commonly supported link. The following diagram provides a conceptual view of an OSI model for peer-to-peer communication:

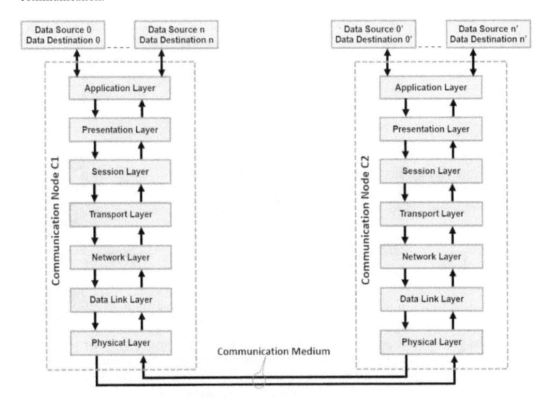

Figure 14.2 – Peer-to-peer communication using the OSI model

In a switched communication topology, the communicating devices connect to an intermediate device or devices called switching nodes, and they exchange data with the switching nodes. The following diagram provides a conceptual view of an OSI model for a switchable communication protocol:

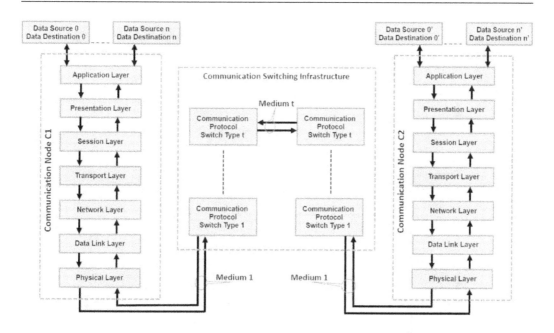

Figure 14.3 – Switchable communication between two nodes using the OSI model

The switching operation may by itself go through many mediums and protocols all abstracted underneath a given layer without the destination node noticing anything about the transformations during the lifetime of the data transfer. This is a huge advantage of following the OSI model recommendation, so whatever transformation the information (packets of data) goes through, the delivered end data to the destination node arrives exactly as it left the initial sending node. Many techniques are used, of course, to guarantee data integrity as it flows from layer to layer and from device to device in the data transit. It is worth noting that not all communication protocols are switchable. However, with the advances in technology, and specifically, the speed at which data can be transferred, protocols are sometimes transported over other protocols that are switchable.

Example communication protocols and mapping to the OSI model

It is probably hard to come up with a communication protocol that offers all the OSI layers in one stack. However, we find that if two or more protocols are grouped together, they can provide a complete mapping of the OSI model. An example is the **Transmission Control Protocol** and **Internet Protocol** (**TCP/IP**), which requires a few more lower layers to make it suitable for communication between two or more physical nodes. The following figure provides a possible mapping between the combination of the TCP/IP suite and the Ethernet protocols to form a full communication stack mapped to the OSI model:

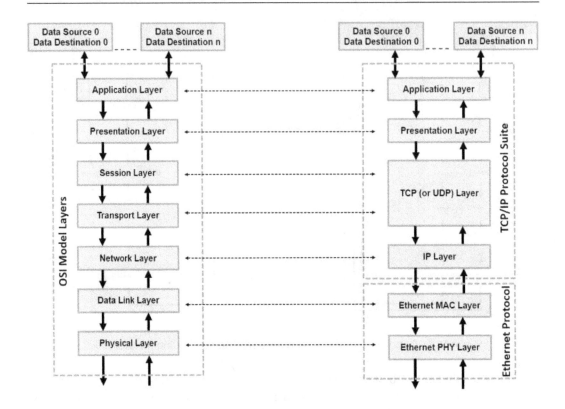

Figure 14.4 – Association of the TCP/IP suite and Ethernet protocols mapping to the OSI model

In this union of protocols association of the TCP/IP suite and the Ethernet protocol, we can achieve a full communication stack capability with the possibility of exchanging data between two physical nodes. The physical nodes can be connected on a peer-to-peer topology or via a switched network that can by itself be of many types.

Communication protocol layers mapping onto FPGA-based SoCs

In the last example, where the TCP/IP protocol suite was associated with the Ethernet protocol to provide a full OSI model communication stack, another mapping can be performed, but this time to implement the model on a real solution. If we take the Zynq-7000 SoC FPGA, we can have the possible mappings illustrated in the following figure:

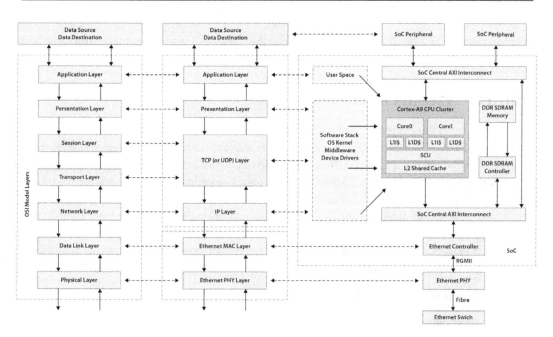

Figure 14.5 – Mapping of the full communication stack to the Zynq-7000 SoC

In this mapping, the source of the data to transmit from the SoC and the destination of the data received from the communication stack can be the SoC peripherals. In a **Closed-Circuit Television** (**CCTV**) application, data in the form of a video stream can be collected from a digital camera, processed, and then transmitted over TCP/IP. The SoC is connected to the **Local Area Network** (**LAN**) to which the security central command room is also connected with control computers and digital display units. The video stream collected by the security camera is fed to the SoC via a video interface. It is then digitally processed within the SoC and only data of interest is streamed over the network. The video data is passed from the camera to a user application where some algorithmic processing is performed on it. Once ready, the data is then packetized and passed to a kernel module performing encryption and compression on it either in software or using the **Programmable Logic** (**PL**) accelerators. The kernel module then calls the OS networking services over a TCP/IP connection when data packets are ready to transmit. When the OS schedules the data for connection, it calls the middleware stack to perform the data transfer on behalf of the initial user application. Data is processed serially by the TCP layer and then passed over to the IP layer, which adds the destination information and then puts it in a queue. It then sends the request to the OS to perform another call to the Ethernet drivers to pass the data in as an Ethernet frame to the Ethernet MAC. The Ethernet MAC serializes the data, performs all the required operations on it, and then streams it over the RGMII interface to the Ethernet PHY. The Ethernet PHY is just a medium translator that takes logical information and transforms it into lights to be transmitted over fiber to the Ethernet LAN switch. When data is received from the network and its destination is the SoC, a symmetrical mechanism is performed on it starting from the Ethernet PHY all the way to the destination user application that will consume it as programmed to do so. Data received by the Zynq-7000 SoC FPGA from the security central command via the Ethernet interface and the TCP/IP stack running on the Cortex-A9 can be a motor control command for camera tracking, image filters update information, and so on.

Control systems overview

Control systems in this context refer to the industrial control systems known as **Programmable Logic Controllers (PLCs)**. PLCs classically performed a set of control operations by actioning certain responses because of a captured input stimulus. They are industrial computing engines packaged appropriately to operate in harsh environments. They are used to control assembly lines in manufacturing, robots, general machinery, and high-reliability operations. They have a relatively simple programming interface and a high capability of detecting and dealing with operational faults. PLCs have been around since the 1960s when they were first introduced in the automobile industry in the United States. From an architecture perspective, a PLC is simply a microprocessor-based controller for industrial applications with storage memory for hosting the executable application to run on it. A PLC includes the following:

- A **Central Processing Unit (CPU)**

- A **Power Supply Unit (PSU)**

- A **Memory Unit (MU)**

- **Input and Output (IO)** Interfaces

- **Communications Interfaces (CIs)**

The CPU receives input from its operating environment via a set of interfaces. It performs on the input the necessary computational operations as indicated by the control program stored in the PLC memory and generates the corresponding results. These results are sent through its output interfaces.

The **Power Supply Unit (PSU)** converts the **Alternate Current (AC)** to a **Direct Current (DC)** power source for the PLC unit.

The MU stores data to operate on from the input interfaces and the program to be executed by the CPU.

The IOs are the interfaces via which the controller communicates with its direct environment, and via which it receives the stimulus and provides the data to the neighboring devices within the operating environment.

The CIs are the remote communication interfaces via which the PLC transmits and receives data, usually through industrial networks such as industrial Ethernet.

Information

For more information about the history of PLCs, please check out the following white paper:
`https://www.c3controls.com/white-paper/history-of-programmable-logic-controllers/`

PLC programming devices are used to develop and download the PLC executable program into the MU of the controller. In general, there is a **Real-Time Operating System** (**RTOS**) running on the PLC CPU and managing the different software threads implementing the PLC control program. The following figure provides an architectural overview of modern PLCs:

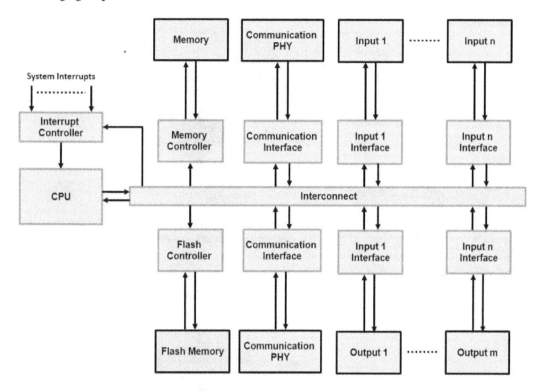

Figure 14.6 – PLC industrial controller architecture overview

With the advance in technology and industry standards, PLCs have also seen an evolution in their role and an expansion into other domains and applications, as we will cover in the next section.

Control system hardware and software mappings onto FPGA-based SoCs

Modern PLCs are intelligent nodes deployed in the Edge and can have versatile and customizable added functionality. PLCs run software that conforms to the IEC 61131 standard, but expanding their functionality into parallel operations, such as image processing, network filtering, and other **Digital Signal Processing** (**DSP**) acceleration, is a desirable added value. The Zynq-7000 SoC provides all the features that are required to build a base PLC industrial controller starting from its PS based on the Cortex-A9 CPU cluster, with its rich peripherals set and the necessary memories and memory interfaces. Its high-throughput interfaces to the associated PL make the functionality expansion an interesting solution in a single chip.

FPGA logic and DSP resources can be used to build computational engines to perform many expansion functions, as well as building custom IOs and communication interfaces that are specific to industrial applications and not available in the PS block. To enhance the PLC based on FPGA SoCs, we can list the following:

- Maintenance operations alerts
- Local DSP algorithm implementation, avoiding data transfers over the network to a central compute node
- Communication protocol acceleration, such as packet filtering and data encryption/decryption
- Fast and deterministic control loop implementation based on hardware state machines
- Image signal processing acceleration
- Software application isolation

It is also worth noting that the design tools used to build the hardware and software for safety-critical operations needs to be certified, which is the case for the Xilinx tools used now for many decades in industrial applications such as PLCs. The following diagram provides a simple mapping for the PLC architecture with the expansion capabilities in FPGA-based SoCs such as the Zynq-7000 SoC:

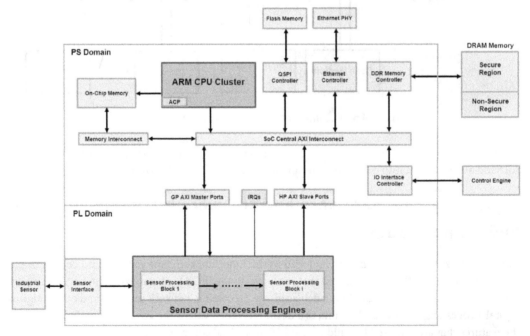

Figure 14.7 – Example extensible PLC controller architecture mapped to the Zynq-7000 SoC

As can be seen in the previous diagram, a PLC based on an FPGA SoC such as the Zynq-7000 SoC has all the required hardware elements, either as hard IPs within the PS or custom IPs to be added from within the PL. This single-chip solution is an attractive option for modern PLCs, which can be implemented with advanced intelligence and computation capabilities.

Summary

In this closing chapter of *Part 3* and this book, we looked at some more advanced applications where FPGA-based SoCs are well suited as a single-chip architecture. They offer a fast time-to-market product development, a lower product cost, and a lower power solution in comparison to multi-chip-based architectures. These advanced applications include sophisticated communication applications that follow the OSI model, or part of it. They also span industrial control applications such as modern PLCs with advanced computation capabilities and feature extensibility using the FPGA PL and DSP computation engines. As we examined in detail in this book, SoCs built using PL have optimal integration and interoperability between the Cortex-A9 CPU cluster and the PL. This flexible architecture offers many capabilities for building communication applications that benefit from the acceleration and filtering extensions that can be hosted in the FPGA logic. PLCs also require more at the Edge processing and intelligence, which is customized to enhance the solution and adapt it to potentially evolving and changing operational environments and conditions.

Part 3 of this book covered some advanced and complex aspects of SoC architecture development and design, including advanced accelerator integration using the Cortex-A9 ACP port and software development using an RTOS. It also addressed the architectural design and system capabilities of video processing, communication protocol implementation, and control system design using FPGA-based SoCs such as the Zynq-7000 SoC devices.

This book covered topics related to SoC architecture design, provided detailed implementation examples targeting the Zynq-7000 SoC FPGA, and made good use of the Xilinx implementation tools. After reading it, if necessary many times, you should be ready to put the skills learned into practice in architecting and building your next SoC, having a high degree of confidence in your product's success.

Good luck!

Questions

Answer the following questions to test your knowledge of this chapter:

1. What is the OSI model? How useful is it if its recommendation is followed when defining a protocol specification?

2. List the OSI model layers and provide a brief description of each layer.

3. What are the main topologies of communication systems? What is the difference between them?

4. To conform to the OSI model, should a communication protocol implement all the OSI model layers?

5. Is TCP/IP a communication protocol that can map directly to the OSI model? Which layers are missing?

6. Which other communication protocol can be appended to the TCP/IP protocol to form a protocol suite that fully implements the OSI model?

7. Which layers of the preceding protocol association can be mapped to the Zynq-7000 SoC FPGA, and which layers need to be implemented using software?

8. What are PLCs and where were they initially used?

9. What are the main elements of a PLC architecture?

10. What are some of the extra functionalities that modern PLCs require to become more attractive and add extra interesting features?

11. How can the FPGA SoC provide the optimal platform to build a modern and intelligent PLC?

Index

M

N

O

Other Books You May Enjoy

If you enjoyed this book, you may be interested in these other books by Packt:

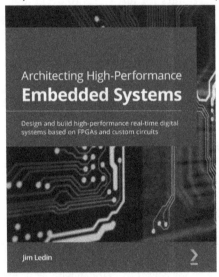

Architecting High-Performance Embedded Systems

Jim Ledin

ISBN: 978-1-78995-596-5

- Understand the fundamentals of real-time embedded systems and sensors
- Discover the capabilities of FPGAs and how to use FPGA development tools
- Learn the principles of digital circuit design and PCB layout with KiCad
- Construct high-speed circuit board prototypes at low cost
- Design and develop high-performance algorithms for FPGAs
- Develop robust, reliable, and efficient firmware in C
- Thoroughly test and debug embedded device hardware and firmware

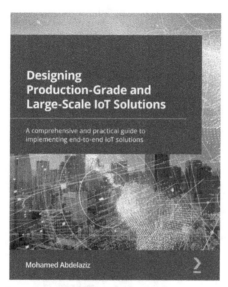

Designing Production-Grade and Large-Scale IoT Solutions

Mohamed Abdelaziz

ISBN: 978-1-83882-925-4

- Understand the detailed anatomy of IoT solutions and explore their building blocks
- Explore IoT connectivity options and protocols used in designing IoT solutions
- Understand the value of IoT platforms in building IoT solutions
- Explore real-time operating systems used in microcontrollers
- Automate device administration tasks with IoT device management
- Master different architecture paradigms and decisions in IoT solutions
- Build and gain insights from IoT analytics solutions
- Get an overview of IoT solution operational excellence pillars

Packt is searching for authors like you

If you're interested in becoming an author for Packt, please visit authors.packtpub.com and apply today. We have worked with thousands of developers and tech professionals, just like you, to help them share their insight with the global tech community. You can make a general application, apply for a specific hot topic that we are recruiting an author for, or submit your own idea.

Share Your Thoughts

Now you've finished *Architecting and Building High Speed SoCs*, we'd love to hear your thoughts! Scan the QR code below to go straight to the Amazon review page for this book and share your feedback or leave a review on the site that you purchased it from.

https://packt.link/r/1801810990

Your review is important to us and the tech community and will help us make sure we're delivering excellent quality content.

Download a free PDF copy of this book

Thanks for purchasing this book!

Do you like to read on the go but are unable to carry your print books everywhere?

Is your eBook purchase not compatible with the device of your choice?

Don't worry, now with every Packt book you get a DRM-free PDF version of that book at no cost.

Read anywhere, any place, on any device. Search, copy, and paste code from your favorite technical books directly into your application.

The perks don't stop there, you can get exclusive access to discounts, newsletters, and great free content in your inbox daily

Follow these simple steps to get the benefits:

1. Scan the QR code or visit the link below

https://packt.link/free-ebook/9781801810999

2. Submit your proof of purchase
3. That's it! We'll send your free PDF and other benefits to your email directly

www.ingramcontent.com/pod-product-compliance
Lightning Source LLC
Chambersburg PA
CBHW081502050326
40690CB00015B/2887